Understanding Transit Accidents Using the National Transit Database and the Role of Transit Intelligent Vehicle Initiative Technology in Reducing Accidents

Final Report

June 2004

Prepared by:

Research and Special Program Administration
Volpe National Transportation Systems Center
Office of Safety and Security
Accident Prevention Division, DTS-73
55 Broadway
Cambridge, MA 02142

Prepared for:

Office of Research, Demonstration, and Innovation
Federal Transit Administration
400 7th Street, SW
Washington, D.C. 20590

and

ITS Joint Program Office
Federal Highway Administration
400 7th Street, SW
Washington, DC 20590

Report Numbers:

FTA-TRI-11-04.1
FHWA-JPO-04-042

Technical Report Documentation Page

1. Report No. FTA-TRI-11-04.1 FHWA-JPO-04-042	2. Government Accession No.	3. Recipient's Catalog No.	
4. Title and Subtitle Understanding Transit Accidents Using the National Transit Database and the Role of Transit Intelligent Vehicle Initiative Technology in Reducing Accidents		5. Report Date June 2004	
		6. Performing Organization Code	
7. Author(s) C. Y. David Yang		8. Performing Organization Report No. DOT-VNTSC-FTA-04-02	
9. Performing Organization Name and Address Volpe National Transportation Systems Center RSPA, US Department of Transportation 55 Broadway Accident Prevention Division, DTS-73 Cambridge, MA 02142		10. Work Unit No. (TRAIS)	
		11. Contract or Grant No.	
12. Sponsoring Agency Name and Address FTA, US DOT 400 7th Street, SW Room 9401 Washington DC 20590	ITS Joint Program Office FHWA, US DOT 400 7th Street, SW Room 3416 Washington DC 20590	13. Type of Report and Period Covered Final	
		14. Sponsoring Agency Code TRI; HOIT-1	
15. Supplementary Notes FHWA COTR – Mr. Raymond J. Resendes, HOIT-1; FTA Task Manager – Brian P. Cronin, TRI-11			

16. Abstract

This report documents the results of bus accident data analysis using the 2002 National Transit Database (NTD) and discusses the potential of using advanced technology being studied and developed under the U.S. Department of Transportation's (U.S. DOT) Intelligent Vehicle Initiative (IVI) program to reduce bus accidents.

The impact of a reduced bus accident rate goes beyond monetary savings. If the number of bus collisions can be effectively reduced using the transit IVI technologies, then:

- Injuries and fatalities from bus accidents will also decrease;
- Transit operators will feel less stress and have more confidence to drive buses in and around congested urban environments;
- Traffic congestion and delays caused by bus accidents will be alleviated; and
- The public will view the bus as a safer mode of travel that is equipped with cutting-edge technology, thereby promoting transit's image, growth, and ridership.

The NTD has important statistics that show vital trends about the transit industry. The "Safety and Security Module" within the NTD contains data regarding incidents reported by transit agencies. Incident records from the Safety and Security Module of the NTD offer useful information to help the U.S. DOT's IVI program develop new technologies to apply on future buses. Currently, several transit IVI projects are developing and testing collision warning systems to assist bus operators in preventing bus accidents by providing them effective and timely warnings. To implement successful transit collision warning systems, a thorough understanding of transit accident types as well as causes and costs of transit accidents are essential.

17. Key Word National Transit Database, Transit Incidents and Cost, Transit Intelligent Vehicle Initiative, Collision Warning Systems		18. Distribution Statement No restrictions.		
19. Security Classif. (of this report) Unclassified	20. Security Classif. (of this page) Unclassified		21. No. of Pages 60	22. Price

Form DOT F 1700.7 (8-72)   Reproduction of completed page authorized

PREFACE

The work presented in this report was conducted by the Accident Prevention Division at the Volpe National Transportation Systmes Center, Research and Special Program Administration, U.S. Department of Transportation (U.S. DOT). This study is sponsored by the Federal Transit Administration (FTA) and the ITS Joint Program Office at the U.S. DOT as part of the Intelligent Vehicle Inititative (IVI) program. Results generated from this study will be an important reference for the transit IVI project teams with their ongoing work of studying and developing transit IVI technologies.

The author would like to thank Mr. Brian P. Cornin of FTA's Office of Research and Technology for his support and guidance during the preparation of this report. Mr. Cronin is the transit platform technical director for U.S. DOT's IVI program. The author would also like to acknowledge Mr. Gerald M. Powers of the Volpe National Transportation Systmes Center for reviewing the bus incident statistics presented in this report.

TABLE OF CONTENTS

Section	Page
EXECUTIVE SUMMARY	xi
1. INTRODUCTION	1
1.1 Background and Motivation of this Study	1
1.2 Organization of the Report	2
2. FINDINGS FROM THE NATIONAL TRANSIT DATABASE (NTD)	3
2.1 Overview of the National Transit Database	3
2.2 General Transit Incident Trends According to the National Transit Database	3
2.3 Major Changes in the 2002 National Transit Database	6
2.4 Summary of Major Incidents from the 2002 National Transit Database	8
2.4.1 General Information	8
2.4.2 Breakdown of Incident Types and Associated Costs	8
2.4.3 Effect of Various Characteristics on Transit Accidents	15
2.5 Summary of Non-Major Incidents from the 2002 National Transit Database	22
2.5.1 General Information	22
2.5.2 Breakdown of Incident Types and Associated Costs	23
2.5.3 Other Information from the Non-Major Incident Records	25
3. COMPARING RESULTS FROM THE NATIONAL TRANSIT DATABASE TO A TRANSIT ACCIDENT ANALYSIS STUDY CONDUCTED BY PATH	27
3.1 Overview of PATH's Transit Accident Data Analysis Work	27
3.1.1 Background on PATH's Project	27
3.1.2 Sources of Data for this Study	27
3.2 Key Results Reported from PATH's Transit Accident Data Analysis	28
4. UTILIZING TRANSIT INTELLIGENT VEHICLE INITIATIVE TECHNOLOGY TO REDUCE TRANSIT BUS ACCIDENTS	31
4.1 Overview of Transit Intelligent Vehicle Initiative Technology Being Developed	31
4.1.1 Frontal Collision Warning System	31
4.1.2 Side Object Detection System and Side Collision Warning System	32
4.1.3 Rear Impact Collision Warning System	33
4.1.4 Integrated Collision Warning System	33
4.2 Potential of Transit Intelligent Vehicle Initiative Technology	34
4.3 Other Bus Accident Costs Besides the Property Damage Cost	39
5. CONCLUSION	41
5.1 Foreseeable Impact of Transit Intelligent Vehicle Initiative Technology	41
5.2 Suggestions for Implementing the "Suitable" Advanced Technology for Your Transit Fleet	41

TABLE OF CONTENTS (Cont.)

Section	**Page**
REFERENCES	43
APPENDIX A. RELATIONSHIP PLOTS ON NUMBER OF COLLISION WARNING SYSTEMS, ANNUAL BUS INCIDENT COUNTS, AND COST RECOVERY TIME FROM SYSTEM DEPLOYMENT	45

LIST OF FIGURES

Figure		Page
2.1.	Number of Bus Collisions from 1991 to 2000	4
2.2.	Bus Accidents from 1991 to 2000, Normalized by Passenger Miles	5
2.3.	Bus-Related Injuries from 1991 to 2000, Normalized by Passenger Miles	5
2.4.	Bus-Related Fatalities from 1991 to 2000, Normalized by Passenger Miles	6
2.5.	A Portion of the Major Incident Reporting Form Used in the NTD [FTA, 2002]	9
2.6.	Collision Detail Sub-Form within the Major Incident Reporting Form [FTA, 2002]	9
2.7.	Collision Types from the Perspective of the Transit Vehicle [FTA, 2002]	10
2.8.	Distribution of Bus Incident Categories from 2002 NTD	11
2.9.	Distribution of Bus Collision Types from the 2002 NTD	12
2.10.	Distribution of Property Damage Costs by Bus Collision Types	13
2.11.	Distribution of Fatalities by Bus Collision Types	13
2.12.	Distribution of Injuries by Bus Collision Types	14
2.13.	Property Damage Cost versus Property Damage Cost Per Bus Collision	15
2.14.	Bus Collision Fatality and Injury Per Bus Collision	16
2.15.	Distribution of Bus Collisions by Intersection Control Types	18
2.16.	Distribution of Bus Collisions by Weather Conditions	19
2.17.	Distribution of Bus Collisions by Lighting Conditions	20
2.18.	Distribution of Bus Collisions by Roadway Conditions	20
2.19.	Distribution of Bus Collisions by Roadway Configurations	21
2.20.	Distribution of Bus Collisions by Roadway Types	22
2.21.	Non-Major Summary Report Form from the NTD [FTA, 2002]	23
2.22.	Distribution of Non-Major Bus Incidents from the 2002 NTD	24
2.23.	Average Property Damage Cost Per Types of Non-Major Safety Incidents	25
A.1.	Time Required to Recoup from Implementation of the Frontal Collision Warning System	45
A.2.	Time Required to Recoup from Implementation of the Side Object Detection System	46
A.3.	Time Required to Recoup from Implementation of the Rear Impact Collision Warning System	47

LIST OF TABLES

Table		Page
2.1.	Breakdown of Major Bus Incidents Recorded in the 2002 NTD[A]	11
2.2.	Breakdown of Transit Bus Collisions According to the 2002 NTD[A]	12
2.3.	Breakdown of Service Types by Transit Bus Collision Types	17
2.4.	Breakdown of Bus Collisions During AM and PM Hours	17
2.5.	Breakdown of Non-Major Bus Incidents Recorded in the 2002 NTD[A]	24
2.6.	Distribution of Non-Major Bus Incidents by Type of Service	26
3.1.	Costs of Bus Collision from 35 Transit Agencies in California[A]	28
3.2.	Percentage Distribution for Various Types of Bus Collisions	29
3.3.	Cost Distribution for Various Types of Bus Collisions	30
4.1.	Potential Cost Saving of Deploying Transit Frontal Collision Warning System	36
4.2.	Potential Cost Saving of Deploying Transit Side Object Detection System	37
4.3.	Potential Cost Saving of Deploying Transit Rear Impact Collision Warning System	38

EXECUTIVE SUMMARY

This report documents the results of bus accident data analysis using the 2002 National Transit Database (NTD) and discusses the potential of using advanced technology being studied and developed under the U.S. Department of Transportation's (U.S. DOT) Intelligent Vehicle Initiative (IVI) program to reduce bus accidents.

The impact of a reduced bus accident rate goes beyond monetary savings. If the number of bus collisions can be effectively reduced using the transit IVI technologies, then:

- Injuries and fatalities from bus accidents will also decrease;
- Transit operators will feel less stress and have more confidence to drive buses in and around congested urban environments;
- Traffic congestion and delays caused by bus accidents will be alleviated; and
- The public will view the bus as a safer mode of travel that is equipped with cutting-edge technology, thereby promoting transit's image, growth, and ridership.

The NTD has important statistics that show vital trends about the transit industry. The "Safety and Security Module" within the NTD contains data regarding incidents reported by transit agencies. Incident records from the Safety and Security Module of the NTD offer useful information to help the U.S. DOT's IVI program develop new technologies to apply on buses. Currently, several transit IVI projects are developing and testing collision warning systems to assist bus operators in preventing bus accidents by providing them effective and timely warnings. To implement successful transit collision warning systems, a thorough understanding of transit accident types as well as causes and costs of transit accidents are essential. The Federal Transit Administration (FTA) tasked the Volpe National Transportation Systems Center to conduct a transit accident analysis using the 2002 NTD.

In 2000 and 2001, the FTA reevaluated the structure of the NTD, and redesigned it in 2002 to better serve NTD users and reporters. Fiscal year 2002 was essentially the beta test period for the newly designed NTD, specifically for the Safety and Security Module, where continuous modifications and refinements were taking place based upon inputs from the FTA and transit agencies that use the NTD. The Safety and Security Module of the 2002 NTD includes two forms for reporting incidents such as bus accidents and crimes that occur on transit vehicles: Major Incident Reporting form (S&S-40) and Non-Major Summary Report form (S&S-50).

The Major Incident Reporting form gathers detailed information on the most severe safety and security incidents occurring in the transit environment. Detailed data from sources such as accident and police reports are used to complete the Major Incident Reporting form. One Major Incident Reporting form is completed for each major incident that occurs at an agency.

According to the 2002 NTD, a total of 1,503 bus-related major incident records were filed. In 2002, 149 transit bus providers throughout the United States filed major incident reports. The number of major incidents reported by various transit agencies ranged from as low as 1 major incident to 165 major incidents. Out of the 1,503 major bus incident records, 1,204 were filed under the "Collision" category that resulted in $7,406,344 of property damage, 67 fatalities, and 4,193 injuries. With the data that were available from the 2002 NTD, the average property damage cost for a major bus collision was $6,151.

The Non-Major Summary Report form is designed to gather information on less severe transit incidents and is similar in concept to the NTD forms used in the past. One Non-Major Summary Report form is completed per reporting period.

The 2002 NTD has 21,382 bus-related non-major incident records. A total of 369 transit agencies throughout the United States provided information on non-major bus incidents in 2002. The number of non-major bus incidents in monthly reports submitted by various transit agencies ranged from 0 to 474 non-major incidents. There were 12,450 non-major safety incidents reported in the 2002 NTD, resulting in more than $13 million in property damage. The average property damage cost for a non-major bus collision was $1,731.

The primary objective of transit IVI technologies is to provide bus operators with effective and timely warnings regarding potential accidents. By reducing the number of potential transit accidents with the help of collision warning systems, transit agencies should anticipate significant cost savings. A calculation from hypothetical examples presented in Section 4 of this report showed that if a transit agency plans to install 80 frontal collision warning systems on its bus fleet and is currently experiencing 40 major frontal collisions and 60 non-major collisions per year, the expected annual monetary savings in property damage costs could be in the range of $200,000 to $300,000. If the cost to implement a frontal collision warning system were $7,500, then the time required to recoup from the implementation cost of 80 frontal collision warning system will be 2 years.

Bus accidents usually have other 'tangible" costs besides property damage. Examples of other bus accident costs include:

- Medical costs to treat employees and passengers injured from bus accidents;
- Workforce adjustment costs such as drug tests and job retraining for bus operators involved in accidents, assignment and training of replacement bus operators, and the administrative efforts to process work changes;
- Emergency response costs for police officers, fire fighters, and emergency personnel; and
- Insurance administration and legal/court costs such as effort to process insurance claims, legal fees associated with accident litigation, and payment/settlement fees related to accidents.

Since the actual cost of a bus accident is probably higher than the estimates presented in Section 4, the return-on-investment from implementing transit IVI technologies should be even greater than the examples presented.

1. INTRODUCTION

1.1 Background and Motivation of this Study

The National Transit Database (NTD) holds crucial statistics that reveal important trends about the transit industry. The "Safety and Security Module" within the NTD contains data regarding incidents (e.g., bus collision, passenger injury, and vehicle damage and theft) reported by transit agencies on a monthly (agencies with 100 or more vehicles in operation) or quarterly basis (agencies with less than 100 vehicles in operation and not operating rail or ferryboat).

Incident records from the Safety and Security Module of the NTD offer useful information to help the U.S. Department of Transportation's (U.S. DOT) Intelligent Vehicle Initiative (IVI) program develop new technologies to apply on buses. As part of the U.S. DOT's Intelligent Transportation Systems (ITS) program, the IVI is aimed at accelerating the development and use of driving assistance and control intervention systems to help drivers operate vehicles more safely and, consequently, reduce vehicle accidents. The U.S. DOT's ITS Joint Program Office coordinates all IVI activities. The Federal Transit Administration (FTA) oversees the transit vehicle "platform" of the IVI program.

Currently, several transit IVI projects are developing and testing collision warning systems that will provide effective and timely warnings to assist bus operators in preventing bus accidents. To implement successful transit collision warning systems, a thorough understanding of transit accident types, as well as causes and costs of transit accidents, are essential. Consequently, the FTA and ITS Joint Program Office tasked the Volpe National Transportation Systems Center to conduct a transit accident analysis using the 2002 NTD.

The key objectives of this transit incident analysis using the NTD included:

- Gaining a better understanding of the frequencies, patterns, and types of transit bus accidents;
- Finding significant causal factors for transit bus accidents;
- Estimating the costs associated with different types of bus accidents based on the information available from the database;
- Discussing the potential of applying collision warning systems to reduce various types of bus accidents and estimating potential return-on-investment of these collision warning systems; and
- Providing guidance on the future research direction for transit IVI studies and development.

Findings and recommendations generated from this study will be provided to transit IVI project teams to assist them in ongoing work of studying and developing transit IVI technologies. In addition, results yielded from this accident analysis work will be presented to transit agency representatives to show the "entire picture" of transit accidents, illustrate the potential benefits of transit collision warning systems in reducing accidents, and assist them in choosing the appropriate collision warning technology for deployment.

1.2 Organization of the Report

This report consists of five major sections. Background information and motivation of this study are discussed in the first section. An overview of the NTD and important findings on transit incidents using the data from the 2002 NTD will be presented in Section 2.

Section 3 will compare the results from this study to selected findings from a transit accident analysis conducted by the California PATH Program at the University of California. Section 4 of this report introduces cutting edge research studies in the area of collision warning systems, funded under the Transit IVI program. The collision warning systems have the great potential to reduce transit bus accidents. Finally, Section 5 will address the anticipated impact of Transit IVI technology in reducing transit incidents.

2. FINDINGS FROM THE NATIONAL TRANSIT DATABASE (NTD)

This section presents findings from the NTD. Sections 2.1 and 2.2 provide a 'high-level" summary of the NTD and present some general transit incident trends in accordance with the NTD. Sections 2.3 through 2.5 will discuss the major changes made in reporting data to the 2002 NTD and show key results from such a dataset.

2.1 Overview of the National Transit Database

The NTD is the primary database maintained by the FTA to keep track of all vital statistics (e.g., financial and operating data) in the transit industry. Various FTA programs and projects are managed and funded based on data gathered from the NTD. Transit agencies that receive the Urbanized Area Formula Program (Section 5307) grants are required to submit data to the NTD. Currently, over 600 transit agencies and authorities submit data to the NTD. However, the FTA encourages all public and private providers of mass transit services to report their services to the NTD program. The NTD can be more reflective of the entire transit industry through the submission of complete and accurate data.

The legislative requirement for the NTD is found in Title 49 U.S.C. 5335(a):

Section 5335(a) National Transit Database (1) To help meet the needs of individual mass transportation systems, the United States Government, state and local governments, and the public for information on which to base mass transportation service planning, the Secretary of Transportation shall maintain a reporting system, using uniform categories, to accumulate mass transportation financial and operating information and using a uniform system of accounts. The reporting and uniform systems shall contain appropriate information to help any level of government make a public sector investment decision. The Secretary may request and receive appropriate information from any source.

(a)(2) The Secretary may make a grant under Section 5307 of this title only if the applicant and any person that will receive benefits directly from the grant are subject to the reporting and uniform systems.

2.2 General Transit Incident Trends According to the National Transit Database

This section presents summary transit accident statistics (in figures) extracted from the 1991 to 2000 NTD.[1] These summary statistics are documented in a government report [Powers, March 2002]. Information presented in this section shows long-term (10 years) national trends on bus accidents. These national trends provide important background material that will facilitate the subsequent discussion in this report.

Figure 2.1 shows the number of bus collisions reported by transit agencies. Types of collisions reported in the NTD include those with vehicles, objects, and people. Figure 2.1 reveals an encouraging trend: the number of bus collisions has decreased in a 10-year span from 1991 to 2000. Comparing the records from 1991 versus 2000 revealed a reduction of 50.2 percent in the

---

[1] Transit accident statistics recorded in the NTD from 1991 to 2000 were based on information submitted by transit agencies using the summary based Transit Safety and Security Form (Form 405). Data gathered using Form 405 are relatively limited compared to the existing data collection tool being used for the 2002 NTD.

number of bus collisions. However, reduction was achieved primarily in the period from 1991 to 1995 and has remained stable from 1996 to 2000.

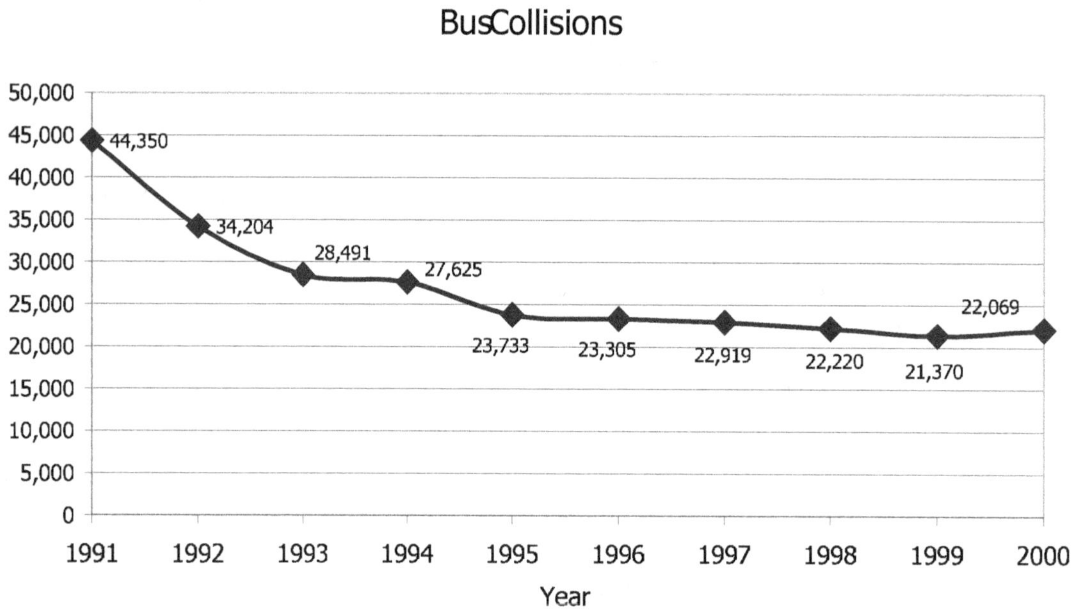

BusCollisions

Year	Collisions
1991	44,350
1992	34,204
1993	28,491
1994	27,625
1995	23,733
1996	23,305
1997	22,919
1998	22,220
1999	21,370
2000	22,069

Figure 2.1. Number of Bus Collisions from 1991 to 2000

Bus accidents from 1991 to 2000, normalized by the annual passenger miles, are presented in Figure 2.2. The plot pattern in Figure 2.2 shows that normalized bus accidents have decreased by 37.9 percent in a 10-year span (1991 value versus 2000 value) but shows no significant accident reduction from 1996 to 2000. This trend is similar to the "bus collisions" plot shown in Figure 2.1.

Bus-related injuries from 1991 to 2000, normalized by the annual passenger miles, are presented graphically in Figure 2.3. Normalized bus fatalities during the same time period are shown in Figure 2.4. Plots on Figures 2.3 and 2.4 clearly indicate that, despite reductions in bus collisions and accidents from 1991 to 2000, injuries and fatalities related to bus incidents have remained fairly steady during this time period.

During the past decade, many transit agencies and authorities have invested significant resources to enhance and promote the safety of bus operation and travel. Training programs have been updated and transit vehicle designs have been improved to better prepare bus operators for the ever-demanding driving environment on urban streets. Promotional materials have been developed to educate bus riders and the general public about safety precautions on and around buses. Accident statistics shown in this section confirm that investments made by transit agencies have improved bus operation safety. However, safety enhancement initiatives introduced previously seem to have reached a plateau since number of bus collisions and accidents have remain fairly stable since the mid-1990s.

To continue to improve safety in bus operation and travel, a better understanding of bus accidents is necessary. Utilizing the abundant data from the NTD will help characterize the causes of transit bus accidents. Consequently, new technologies with the goal of further reducing bus collisions and accidents can be researched and developed. The remaining portion of this report presents findings of a detailed transit accident analysis utilizing the NTD and introduces some of the new transit technologies with great potential in reducing accidents.

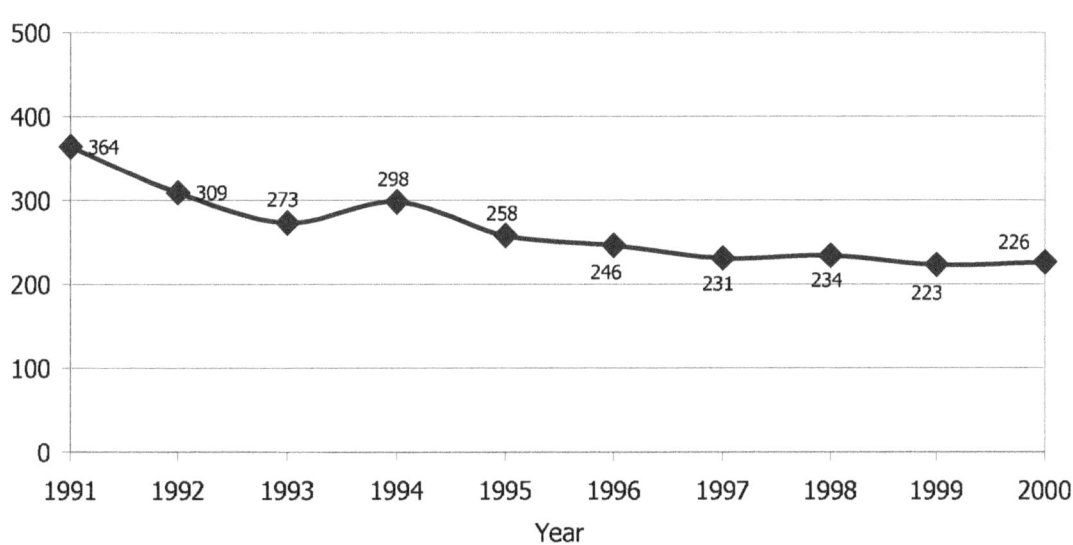

Figure 2.2. Bus Accidents from 1991 to 2000, Normalized by Passenger Miles

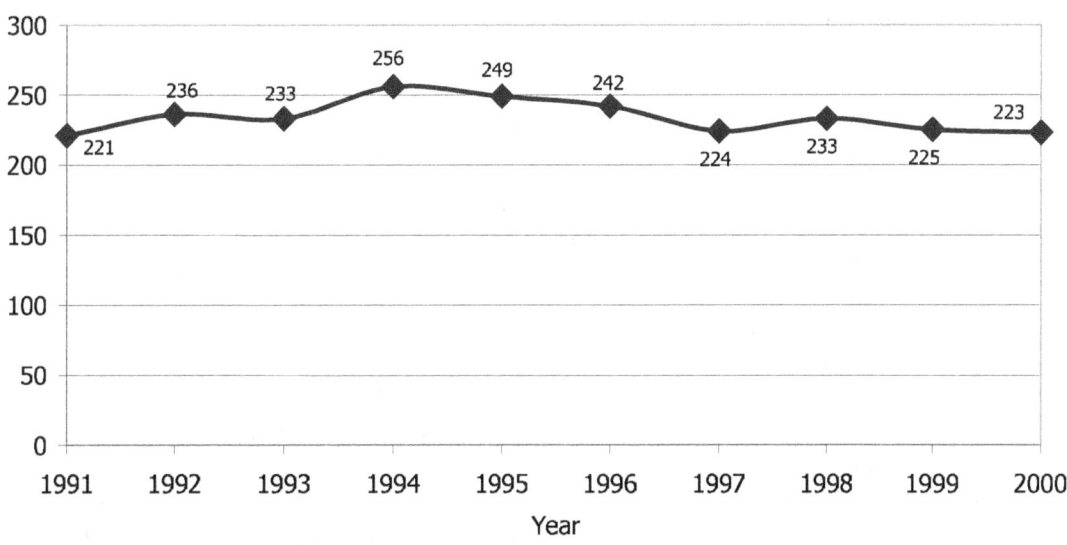

Figure 2.3. Bus-Related Injuries from 1991 to 2000, Normalized by Passenger Miles

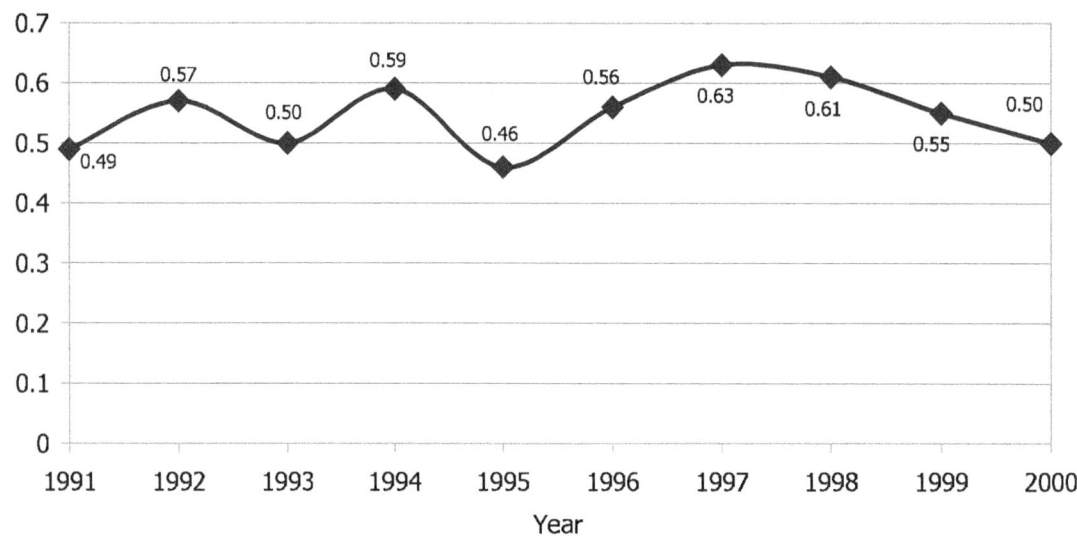

Figure 2.4. Bus-Related Fatalities from 1991 to 2000, Normalized by Passenger Miles

2.3 Major Changes in the 2002 National Transit Database

In 2000 and 2001, the FTA reevaluated the structure of the NTD, involving extensive outreach to the transit industry. This task assessed the usefulness of NTD data to various constituencies, balancing the usefulness of the data with the reporting burden to transit agencies. As a result, the FTA redesigned the NTD in 2002 to better serve NTD users and reporters.

Some of the major changes in the 2002 NTD are presented in this section:

1. Data reporting has been restructured into eight reporting modules, each containing one or more forms to allow transit agencies and authorities to submit the necessary data:
    - Basic Information Module (3 forms – *Transit Agency Identification*, *Transit Agency Contacts*, and *Contractual Relationship*)
    - Financial Module (4 forms – *Sources of Funds*, *Uses of Capital*, *Operating Expenses*, and *Operators' Wages*)
    - Asset Module (3 forms – *Stations and Maintenance Facilities*, *Transit Way Mileage*, and *Revenue Vehicle Inventory*)
    - Transit Agency Service Module (2 forms – *Transit Agency Service* and *Fixed Guideway Segments*)
    - Resource Module (3 forms – *Employees*, *Maintenance Performance*, and *Energy Consumption*)
    - Safety and Security Module (5 forms – *Incident Mode Service*, *Ridership Activity*, *Security Configuration*, *Major Incident Report*, and *Non-Major Summary Report*)
    - Federal Funding Allocation Statistics Module (1 form – *Federal Funding Allocation Statistics*)
    - Preliminary Government Performance and Results Act Summary Module (1 form – *Preliminary Government Performance and Results Act Summary*)

2. A new component has been added to the asset module to enhance reporting of public agency transportation assets and the projected renewal cost. Information provided in this module will help in assessing the condition of transit assets and the need for reinvestment to maintain the assets in good condition.
3. Data reporting to the Safety and Security Module has been enhanced. To record safety and security incidents, a new two-tiered reporting system has replaced the summary-based Transit Safety and Security Form (Form 405) used in previous years. The 2002 NTD collects safety and security data through (1) major incident reporting (captures detailed information on incidents by incident basis), and (2) non-major incident summary reporting. Depending on the fleet size of a transit agency, reporting frequency for safety and security data is either monthly or quarterly.
4. Reporting for capital fund sources (previously the Capital Funding Form) and operations fund sources (previously the Operating Funding Form) is combined into one form in the 2002 NTD. This revised reporting format for earned and expended funds provides a more complete accounting of the transit agency's revenue flow.
5. The software used to report data has various screens tailored to specific modes of transit and types of service. Transit agency reporting is facilitated by customized screens that solicit data on specific mode and service provided by that agency.
6. The 2002 NTD has the automated data validation capability built into the reporting software. This system enhancement provides the transit agencies with better tools to ensure that data reported by them are complete and accurate before submission to FTA.

Changes made in the format of the 2002 NTD include added incident-based reporting for major safety and security incidents and improvements to the interface of the data reporting system, enhanced the quality of the data reported by the transit agencies and authorities, and encouraged additional transit agencies to participate in the NTD program.

Transit agencies use the NTD website (www.ntdprogram.com) to report data. Each transit agency has a unique identification number assigned by FTA to be used in data reporting and corresponding with the NTD. Once the unique identification number is entered by a transit agency via the website, the forms necessary to file that agency's annual report will be generated upon completion of the Incident Mode Service Module in the initial session. (Note: The Incident Mode Service Module establishes the forms necessary for the particular grantee based on the mode, types of service, and number of vehicles reported.)

Fiscal year 2002 was essentially the beta test period for the newly designed NTD, specifically for the Safety and Security Module, where continuous modifications and refinements continue to take` place based upon inputs from the FTA and transit agencies that use NTD. (Note: Design specifications for the Safety and Security Module of the NTD were finalized in December 2001 and the initial Internet-based reporting module was up and running in February 2002.)

Finally, the Safety and Security Module of the 2002 NTD includes two forms for reporting incidents such as bus accidents and crimes that occur on transit vehicles: Major Incident Reporting form (S&S-40) and Non-Major Summary Report form (S&S-50). Sections 2.4 and 2.5 discuss the data gathered from the Major Incident Reporting form and Non-Major Summary Report form respectively.

2.4 Summary of Major Incidents from the 2002 National Transit Database

2.4.1 General Information

The Major Incident Reporting form from the Safety and Security Module of the NTD gathers detailed information on the most severe safety and security incidents occurring in the transit environment. Detailed data from sources such as accident and police reports are used to complete the Major Incident Reporting form. One Major Incident Reporting form is completed for each major incident that occurs at an agency. Major Incident Reporting forms are submitted to the NTD on a monthly basis if total number of operating vehicles in a transit agency is greater than or equal to 100 or quarterly if the agency's operating vehicles number less than 100.

Transit agencies log onto the NTD website to report major incidents. A part of the Major Incident Reporting form used in the NTD is shown in Figure 2.5. A transit incident is considered "major" when one or more of the following conditions occurred:

1. A fatality (excluding suicides or deaths resulting from illnesses);
2. Injuries requiring immediate medical attention away from the scene for two or more persons;
3. Property damage equal to or exceeding $25,000 (for both transit and non-transit vehicles and property);
4. An evacuation of a transit vehicle due to life safety reasons such as fire, fuel leak, and electrical hazard;
5. A collision at a grade crossing resulting in an injury or property damage equal to or exceeding $7,500 (change as of October 2003);
6. A main-line derailment;
7. A collision with person(s) on a rail right-of-way resulting in injuries that require immediate medical attention away from the scene for one or more persons (applies only to rail incidents);
8. A collision between a rail transit vehicle and another rail transit vehicle or a transit non-revenue vehicle resulting in injuries that require immediate medical attention away from the scene for one or more persons (applies only to rail incidents).

2.4.2 Breakdown of Incident Types and Associated Costs

One of the many categories of information gathered by the Major Incident Reporting form is the type of transit incident. Categories of transit incidents available from the Major Incident Reporting form are:

A. Collision
B. Security Incident
C. Derailment
D. Evacuation
E. Fire
F. Vehicle Leaving Roadway
G. Fatality/Injury Not Otherwise Classified (previously personal casualty)

If a reported transit incident is a "Collision," then a "Collision Detail" sub-form (see Figure 2.6) within the Major Incident Reporting form would appear on the NTD website to gather additional information about the collision.

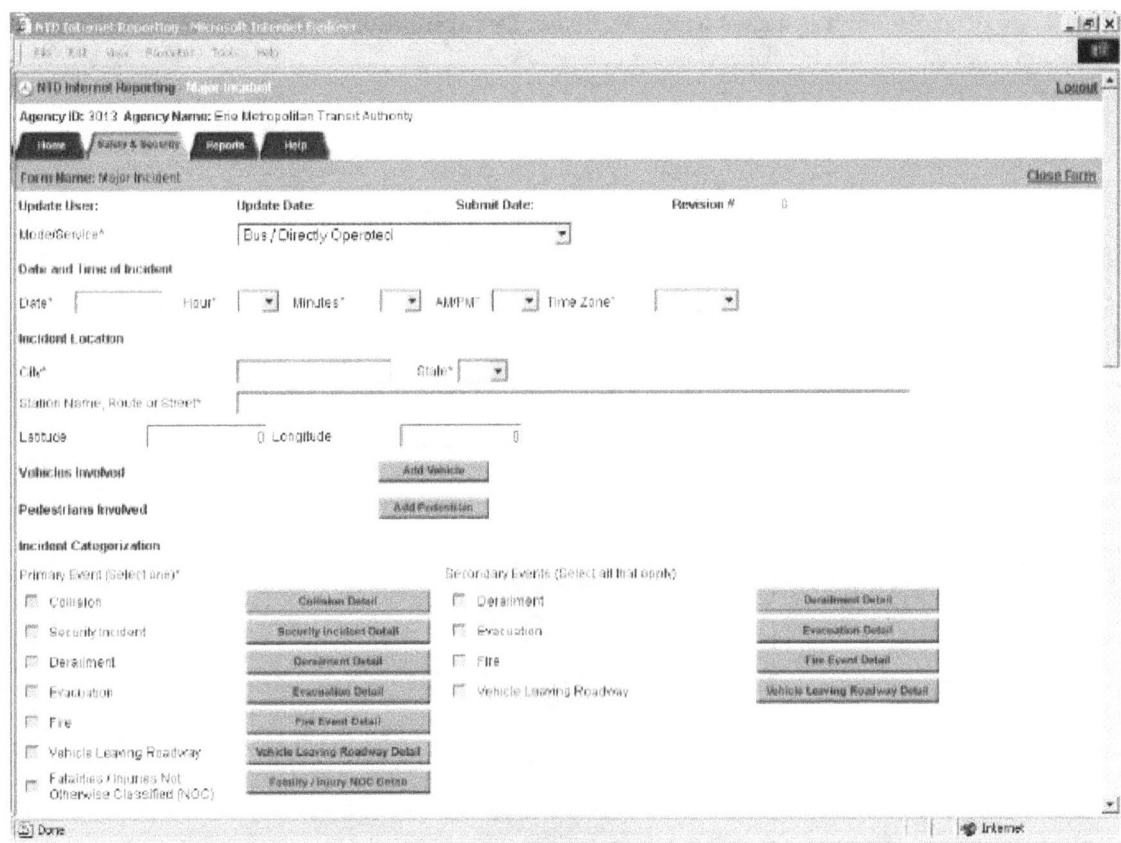

Figure 2.5. A Portion of the Major Incident Reporting Form Used in the NTD [FTA, 2002]

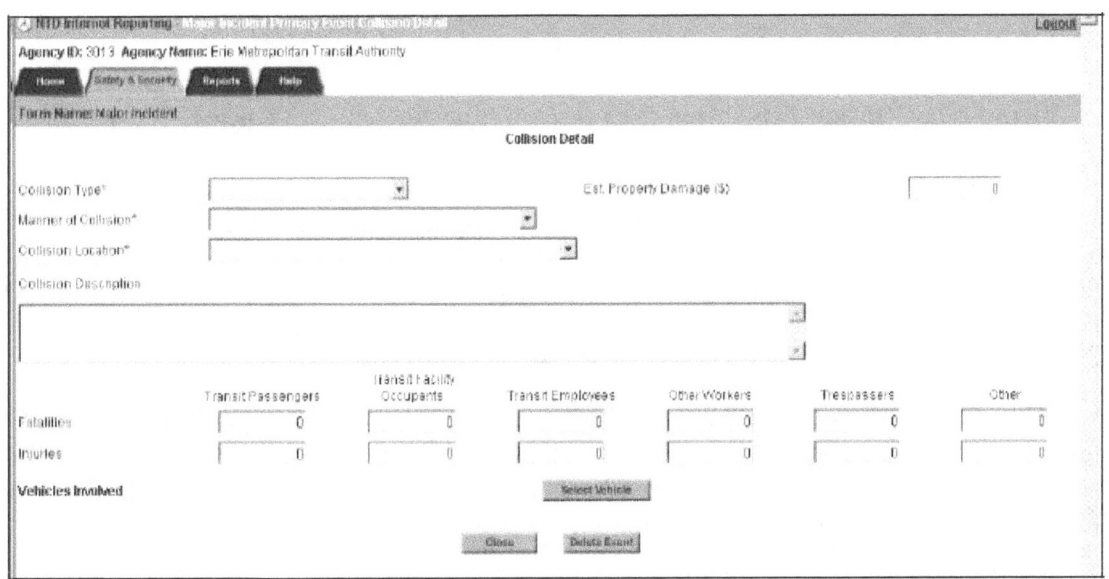

Figure 2.6. Collision Detail Sub-Form within the Major Incident Reporting Form [FTA, 2002]

One piece of information collected by the Collision Detail sub-form is 'Collision Type." Transit collision can be filed into the following six categories:

A. Front
B. Back
C. Angle
D. Sideswipe
E. Fixed Object
F. Other

Figure 2.7 depicts the definitions of front, back, angle, and sideswipe collisions when a transit vehicle collides with other vehicles. Choice of collision type is always from the point of view of the transit vehicle.

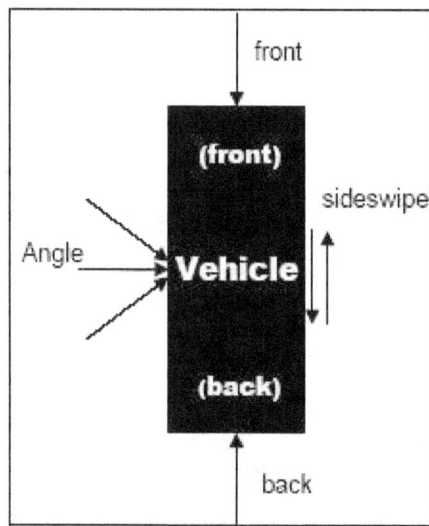

Figure 2.7. Collision Types from the Perspective of the Transit Vehicle [FTA, 2002]

In addition to 'Collision Type," other key information can be gathered from the Collision Detail sub-form including estimated cost in property damage and fatality and injury resulting from the collision. Estimated cost in property damage refers to the dollar amount required to repair/replace all vehicles or public/private property involved in a transit collision. When a transit collision results in death or injury of passengers, transit facility occupants, employees, other workers, or trespassers, the numbers are counted and reported.

According to the 2002 NTD, a total of 1,503 bus-related major incident records were filed and 31 of these records had multiple incident types occur in a single event (e.g., a bus collision followed by a fire). In 2002, 149 transit bus providers throughout the United States filed major incident reports. The number of major incidents reported by various transit agencies ranged from 1 major incident to 165 major incidents.

Out of the 1,503 major bus incident records, 1,204 are filed under the 'Collision" category and 299 records fell under the other 6 transit incident categories (i.e., Security Incident, Derailment, Evacuation, Fire, Vehicle Leaving Roadway, and Fatality/Injury Not Otherwise Classified) (See Table 2.1 and Figure 2.8). The 1,204 bus collisions reported in the 2002 NTD resulted in $7,406,344 of property damage, 67 fatalities, and 4,193 injuries.

Table 2.1. Breakdown of Major Bus Incidents Recorded in the 2002 NTD[A]

Incident Category	Number of Records	Total Property Damage[B]	Number of Fatalities	Number of Injuries
Collision	1,204	$7,406,344	67	4,193
Other (Security Incident, Derailment, Evacuation, Fire, Vehicle Leaving Roadway, and Fatality/Injury Not Otherwise Classified)	299	$3,449,625	14	202
Total	**1,503**	**$10,855,969**	**81**	**4,395**

[A.] Final reporting of the 2002 NTD – information as of July 2, 2003.
[B.] Six transit agencies provided little or no data on property damage cost.

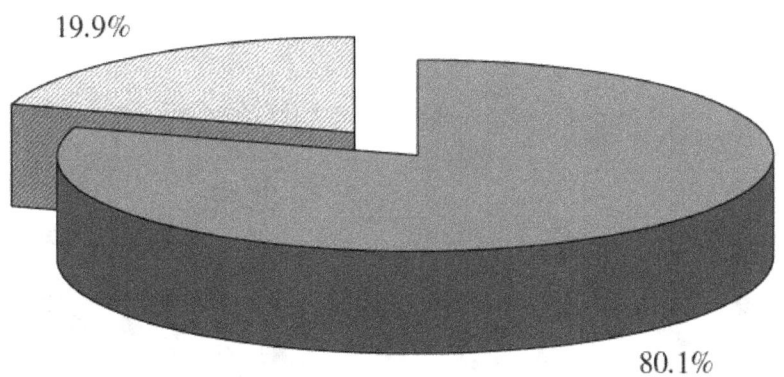

Figure 2.8. Distribution of Bus Incident Categories from 2002 NTD

Further breakdown of the 1,204 records under the "Collision" category revealed additional details about bus collisions. Of the bus accidents recorded in the 2002 NTD, frontal collisions occurred the most often, followed by back collisions, angle collisions, and sideswipes. Table 2.2 provides a summary of various bus collision statistics found in the 2002 NTD. Figures 2.9-2.12 show percentage distributions on the information presented in Table 2.2.

Table 2.2. Breakdown of Transit Bus Collisions According to the 2002 NTD[A]

Type of Bus Collision	Number of Records	Total Property Damage[B]	Number of Fatalities	Number of Injuries
Front	389	$2,439,715	31	1,471
Sideswipe	128	$571,998	1	361
Angle	237	$1,470,848	16	839
Back	353	$1,717,099	9	1,259
Fixed Object/Other	97	$1,206,684	10	263
Total	**1,204**	**$7,406,344**	**67**	**4,193**

[A.] Final reporting of the 2002 NTD – information as of July 2, 2003.
[B.] Six transit agencies provided little or no data on property damage cost.

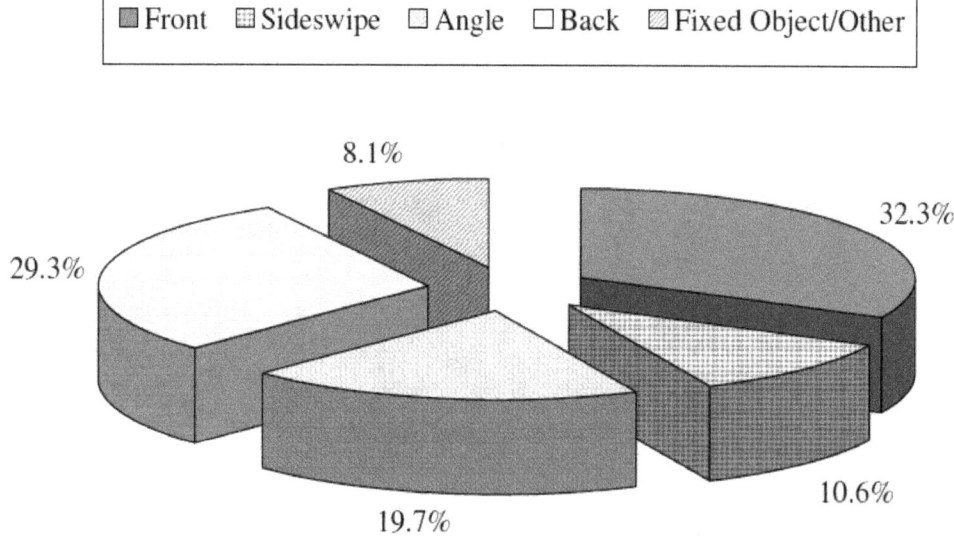

Figure 2.9. Distribution of Bus Collision Types from the 2002 NTD

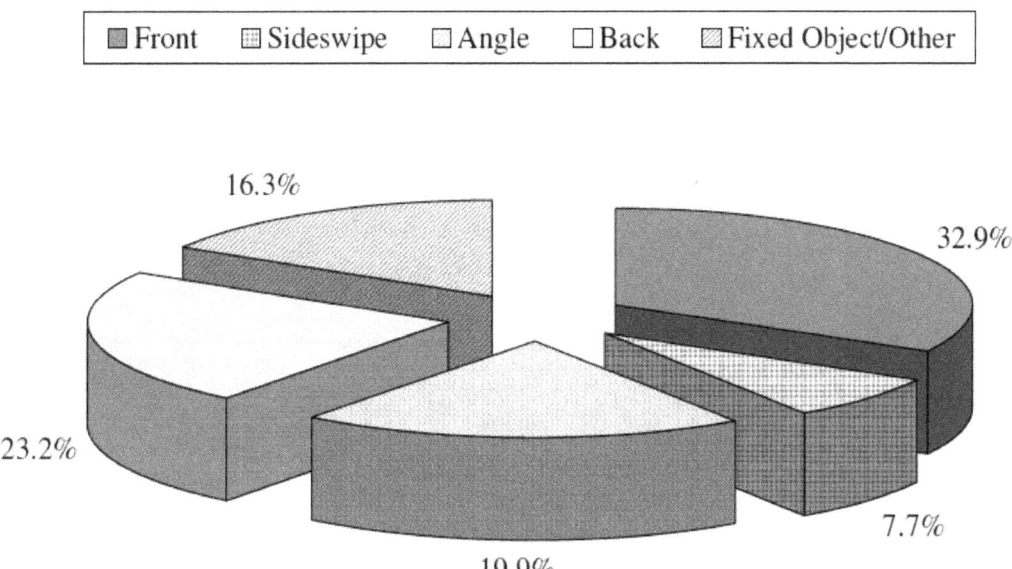

Figure 2.10. Distribution of Property Damage Costs by Bus Collision Types

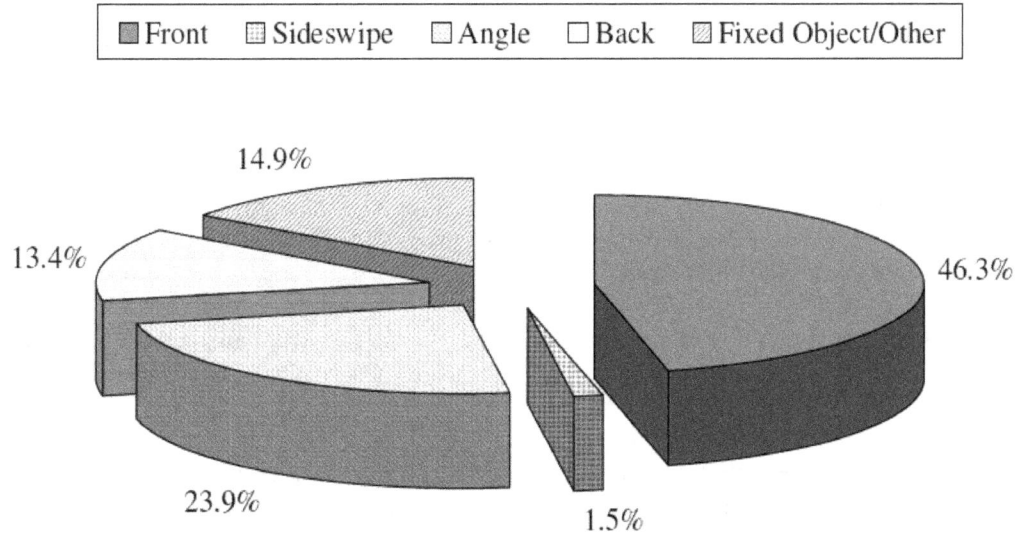

Figure 2.11. Distribution of Fatalities by Bus Collision Types

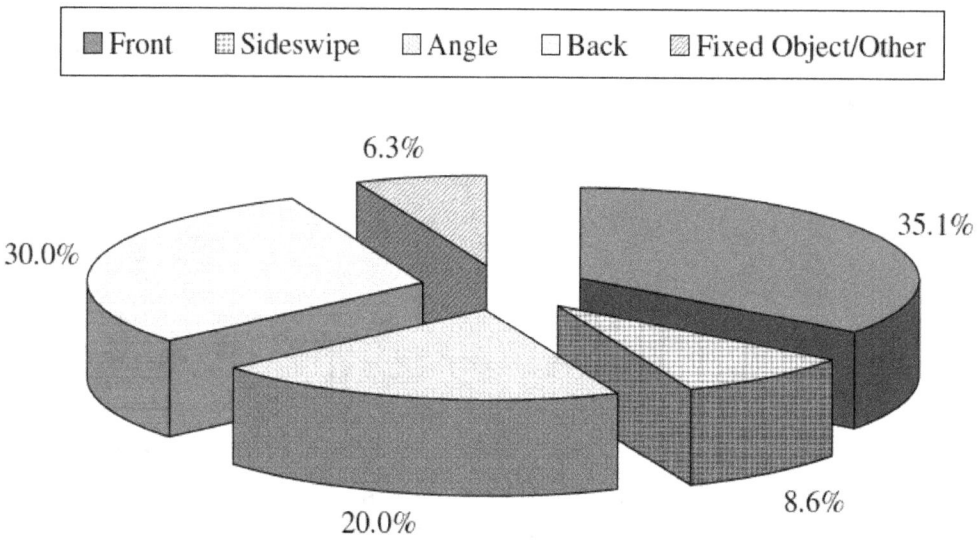

Figure 2.12. Distribution of Injuries by Bus Collision Types

Figures 2.9 and 2.10 show a similar trend for four types of bus collision, where frontal collisions resulted in highest total property damage cost according to the 2002 NTD, followed by back collisions, angle collisions, and sideswipes. Fatality distribution in Figure 2.11 shows that frontal collisions and sideswipes resulted in the highest and lowest number of fatalities respectively. However, Figure 2.12 indicates that frontal and rear bus collisions caused the highest proportions of injuries compared to angle collisions, sideswipes, and collisions with fixed object/other.

To better show the relationship between types of bus collisions versus property damage cost, fatalities, and injuries, values presented in the "Total Property Damage," "Number of Fatalities," and "Number of Injuries" columns from Table 2.2 were normalized using the collision records corresponding to each type of bus collision.

Figure 2.13 compares property damage cost (by type) versus property damage cost normalized by total numbers of collisions (by type). Several interesting observations can be made from Figure 2.13:

1. Property damage cost rate for frontal collision is 1.40 times, 1.01 times, and 1.29 times more than the cost rates for sideswipe, angle collision, and rear collision respectively;
2. Property damage cost rate for angle collision is very close to the cost rate for frontal collision even though the total property damage cost for frontal collision is almost twice the cost for angle collision;
3. Property damage cost rate for back collision is close to the cost rate for sideswipe, however, the total property damage cost for back collision is three times the cost for sideswipe;
4. Property damage rate for collisions with the back of buses is the second lowest of the five categories, but the total property damage cost for rear collision is the second highest; and

5. Property damage cost per collision for the 'Fixed Object/Other" category is considerably higher compared to the other four categories.

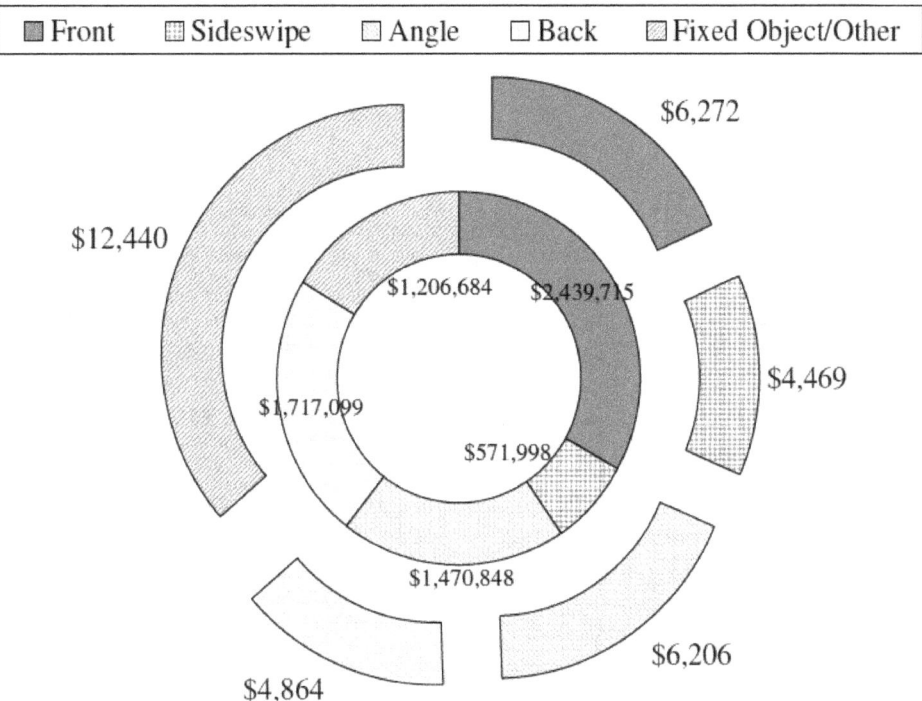

Figure 2.13. Property Damage Cost versus Property Damage Cost Per Bus Collision

Figure 2.14 presents fatality and injury rates (i.e., fatality and injury counts normalized by number of collision records) resulted from various types of bus collisions. Several noteworthy observations are revealed from Figure 2.14:

1. Fatality rates for front, angle, and back collisions are similar, however, injury rate caused by back collisions is considerably lower compared to the injury rates for front and angle collisions;
2. Bus sideswipes yielded second lowest fatality rate and lowest injury rate – such collision causes less impact on buses at the point of contact compared to front, angle, and back collisions; and
3. Injury rate for collisions under the 'Fixed Object/Other" cate gory is highest amongst the five collision types – possibly because more people outside of the bus such as transit commuters waiting at a bus stop or bystanders at an intersection are involved in this collision type.

2.4.3 *Effect of Various Characteristics on Transit Accidents*

This subsection presents additional information regarding major bus incidents reported in the 2002 NTD – the effects of selected environmental and roadway characteristics on bus collisions are examined and discussed.

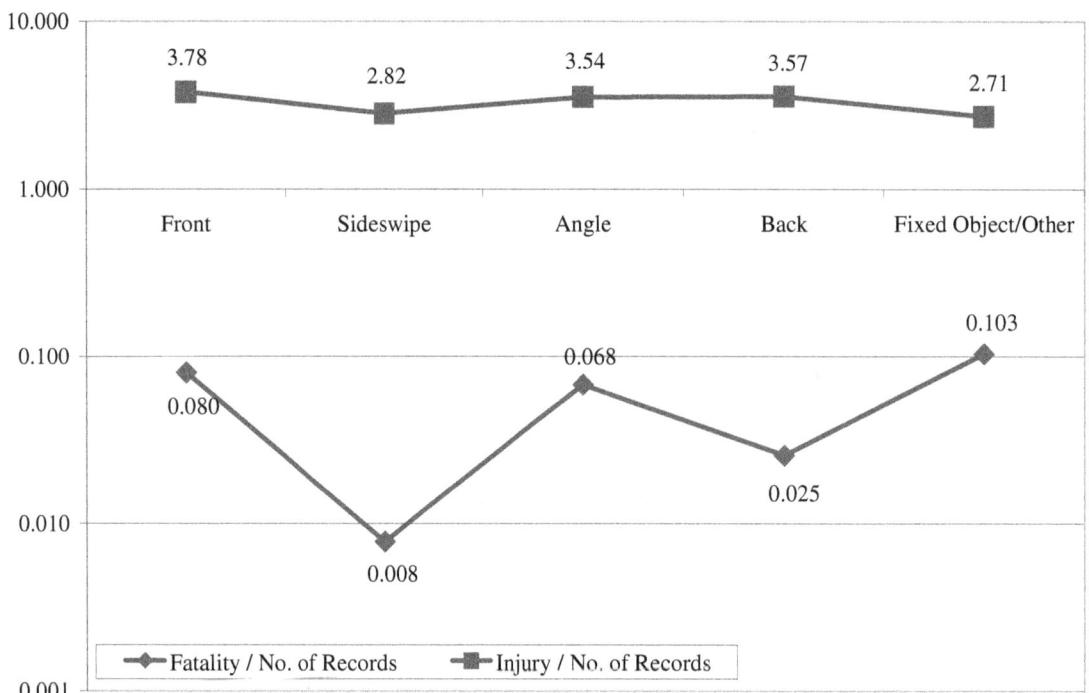

Figure 2.14. Bus Collision Fatality and Injury Per Bus Collision

Factor 1 – Type of Service (Directly Operated versus Purchased Transportation). When transit agencies report major incidents to the 2002 NTD, one piece of information that needs to be specified at the beginning is "service type" of the transit vehicle involved in incidents. Two service types are available for selection:

- *Directly operated service*: transportation service provided directly by a transit agency, using their employees to supply the necessary labor to operate the revenue vehicles;
- *Purchased transportation service*: service provided to a public transit agency from a public or private transportation provider based on a written contract and uses its own employees to operate revenue vehicles (does not include franchising, licensing operations, management services, cooperative agreements, or private conventional bus service).

Of the 1,204 reported bus collisions, 1,135 (94.3 percent) collisions involved directly operated service buses and only 69 (5.7 percent) involved vehicles operated by purchased transportation service. On the national basis, the numbers of buses under the "directly operated service" category are considerably higher than buses in the "purchased transportation service" category; consequently, the results reported above are expected. Table 2.3 presents further breakdown of the directly operated service versus purchased transportation service by types of collision.

Table 2.3. Breakdown of Service Types by Transit Bus Collision Types

Type of Bus Collision	Records in Directly Operated Service	Records in Purchased Transportation Service
Front	374	15
Sideswipe	116	11
Angle	229	8
Back	322	31
Fixed Object/Other	94	4
Total	**1,135**	**69**

Factor 2 – Time of Day (AM versus PM). Of the 1,204 bus collisions filed as major incidents in the 2002 NTD, 447 (37.1 percent) collisions occurred during the AM hours (12:00 AM to 11:00 AM) and 757 (62.9 percent) happened in the PM hours (12:00 PM to 11:00 PM). The core hours of bus operation usually range from 6:00 AM to 8:00 PM. Since more core hours fall in the PM time period (12:00 PM to 8:00 PM) compared to the AM time period (6:00 AM to 11:00 AM), it is expected that more bus accidents would fall in the PM category. Results reported above, from the 2002 NTD, are consistent with such expectation. Table 2.4 presents further breakdown of bus collisions occurring during AM and PM hours, by types of collision.

Table 2.4. Breakdown of Bus Collisions During AM and PM Hours

Type of Bus Collision	Collisions During AM Hours	Collisions During PM Hours
Front	144	245
Sideswipe	42	86
Angle	92	145
Back	136	217
Fixed Object/Other	33	64
Total	**447**	**757**

Factor 3 – Intersection Control. This factor examines the correlation between intersection control and bus collision. Major incident reports for the 2002 NTD allow reporting agencies to specify traffic control device in use at/near the incident site (355 out of the 1,204 bus collision records have information on 'intersection control'). Data analysis for this study divided the intersection controls into four categories: no control device, stop sign, traffic signal, and others (yield sign, other signs, police officer/flagman, and crossing gate).

A percentage distribution of bus collision by intersection control type is shown in Figure 2.15. Bus collisions occurred at/near traffic signal is clearly the dominating category at 62.5 percent. Percentage distribution becomes 80.0 percent when combining bus collisions occurred at/near intersections controlled by either traffic signal or stop sign. One possible reason for the high percentage of bus collisions occurring at/near traffic signal and stop sign: many bus routes are located in urban areas where a majority of intersections are controlled by either traffic signals or stop signs.

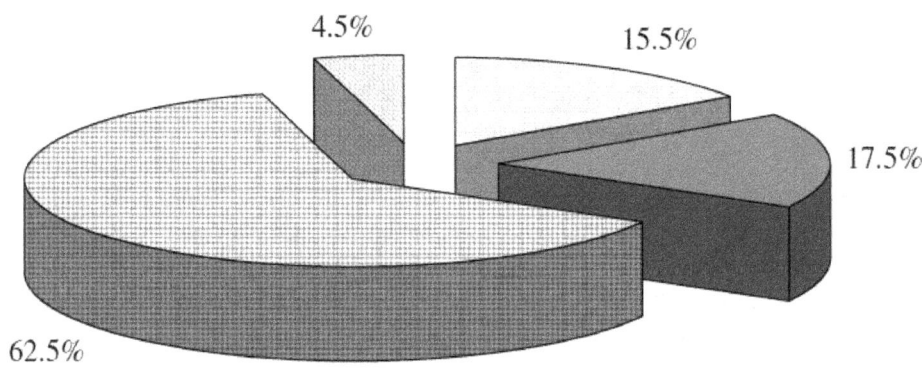

Figure 2.15. Distribution of Bus Collisions by Intersection Control Types

Factor 4 – Weather. Of the 1,204 bus collision records in the 2002 NTD, 884 records have information on the weather condition at the time of accident. Six categories of weather conditions were used in the data analysis: (1) clear, (2) fog or mist, (3) cloudy, (4) rain, (5) snow or sleet, and (6) other.

Figure 2.16 shows the percentage distribution of bus collisions occurring under various weather conditions. According to Figure 2.16, more than three-quarters of bus collisions occurred during clear weather conditions and the number of bus accidents that occurred during adverse weather conditions is relatively small in comparison. The trend presented in Figure 2.16 is consistent with the crash statistics, for all vehicle types, found in the National Highway Traffic Safety Administration's (NHTSA) Fatality Analysis Reporting System (FARS) and General Estimates System (GES) databases: the majority of crashes (> 80 percent) occurred at "normal" weather condition compared to other weather conditions [NHTSA, December 2001; NHTSA, December 2002].

Factor 5 – Lighting. This variable describes the lighting situation at the time of bus accident. Of the 1,204 bus collision records filed as major incidents in the 2002 NTD, 890 records have information on the lighting condition. Transit agencies providing response for this variable have four choices on the 2002 NTD website: (1) daylight, (2) dawn or dust, (3) dark with street lights, and (4) dark with no street lights.

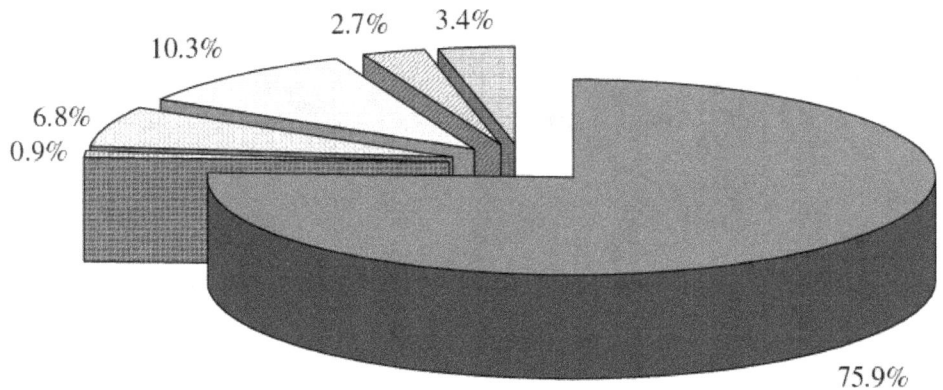

Figure 2.16. Distribution of Bus Collisions by Weather Conditions

Figure 2.17 presents the percentage distribution of bus collisions occurring under the four lighting conditions. Figure 2.17 clearly indicates that bus collisions predominately occurred in conditions where "light" is not a concern (total percent for bus collisions occurring under "daylight" and "dark with street lights" conditions equals 93.6 percent). FARS and GES da ta showed a similar trend where more than 80 percent of crashes (all vehicle types) occurred when the lighting condition is either "daylight" or "dark, but lighted" [NHTSA, December 2001; NHTSA, December 2002].

Factor 6 – Roadway Conditions. The effect of roadway conditions on bus collisions is examined in this factor. Roadway conditions were divided into five categories for the data analysis: (1) dry, (2) wet, (3) snow or slush, (4) ice, and (5) other. Out of the 1,204 bus collision records, 873 have information on roadway conditions at the time of bus accidents.

According to Figure 2.18, only a very small portion of major bus collisions occurred on undesirable roadway conditions such as snow or ice. More than 80 percent of collisions happened when the road is dry.

Results shown in Figure 2.18 are consistent with results for several other factors presented above, suggesting that the probability of a major bus collision occurring in "normal" driving conditions with clear weather, ample lighting, and dry roads is considerably higher compared to "abnormal" driving conditions.

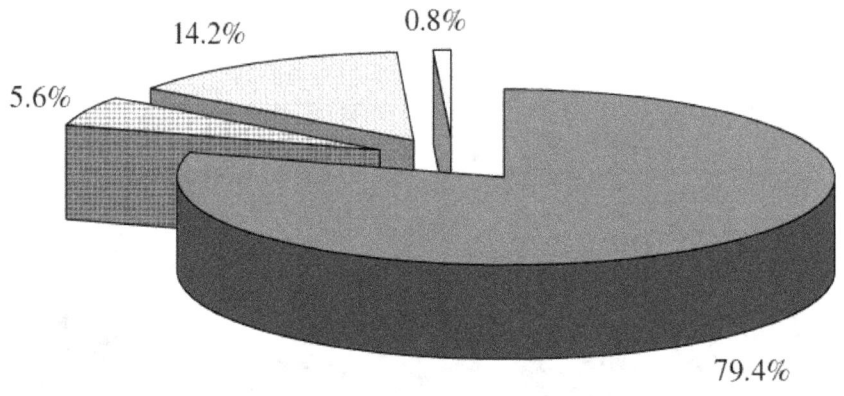

Figure 2.17. Distribution of Bus Collisions by Lighting Conditions

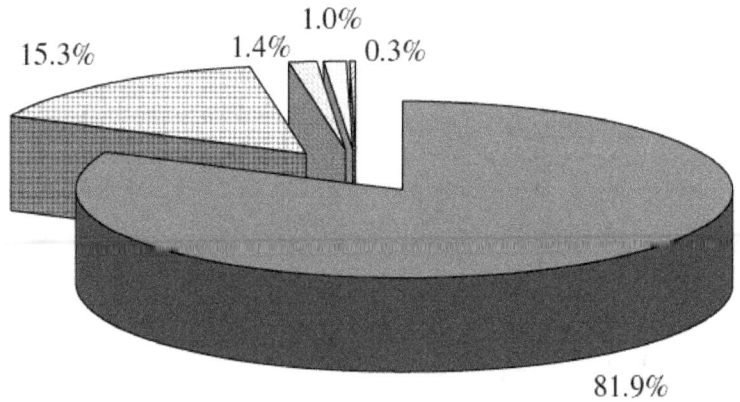

Figure 2.18. Distribution of Bus Collisions by Roadway Conditions

Factor 7 – Roadway Configuration. The relationship between roadway configuration and major bus collision is examined using this factor. The 2002 NTD website offers five choices for the "roadway configuration" factor: (1) straight, (2) level, (3) curve, (4) uphill, and (5) downhill. Of the 1,204 bus collision records filed as major incidents in the 2002 NTD, 864 records have information on this variable.

Figure 2.19 presents the distribution of bus collisions occurring on the five roadway configurations. Once again, most bus collisions took place in a "normal" driving environment – nearly 90 percent of the collisions reported in the 2002 NTD happened on straight roadways.

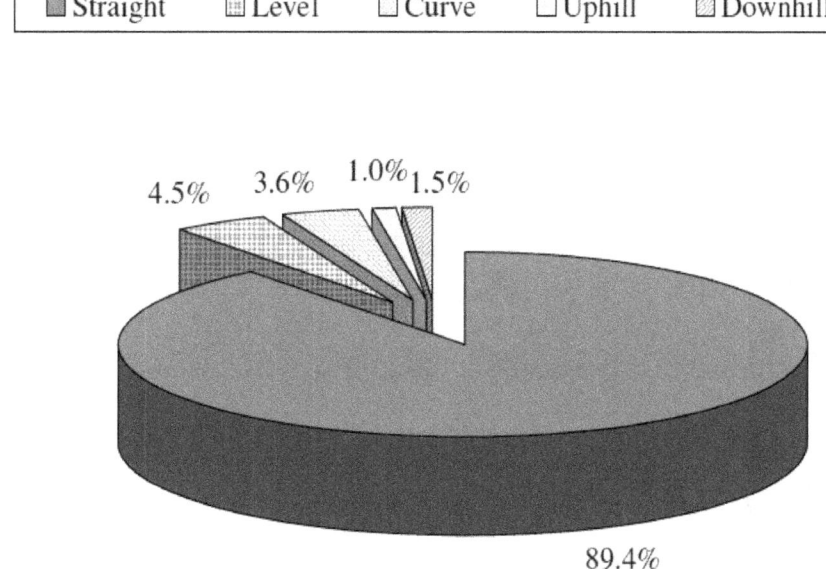

Figure 2.19. Distribution of Bus Collisions by Roadway Configurations

Factor 8 – Roadway Type. Finally, distribution of bus collisions on various roadway types is presented in Figure 2.20. Roadway types were divided into six categories for the data analysis: (1) intersection or grade crossing, (2) divided highway, (3) ramp, (4) bridge, (5) tunnel, and (6) private property. There are 764 major bus accident records that have information on "roadway type."

Figure 2.20 shows that more bus collisions occurred at intersections/grade crossings or divided highways than other roadway types combined. Such results are consistent with expectations because a majority of bus operations occurs in and around urban environments predominated by intersections and divided highways.

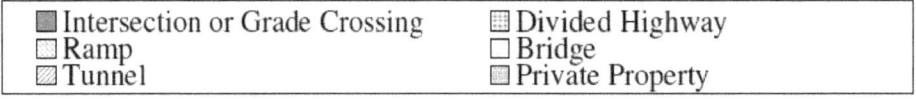

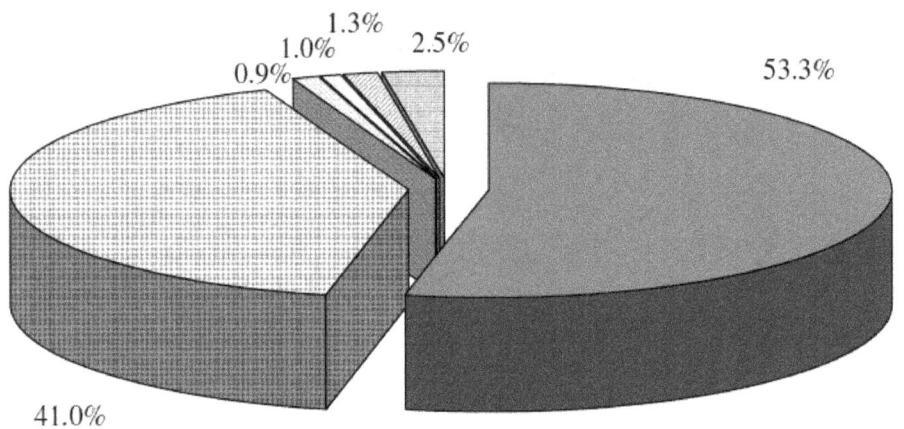

Figure 2.20. Distribution of Bus Collisions by Roadway Types

2.5 Summary of Non-Major Incidents from the 2002 National Transit Database

2.5.1 *General Information*

The Non-Major Summary Report form is designed to gather information on less severe transit incidents and is similar in concept to the NTD forms used in the past. One Non-Major Summary Report is completed per reporting period. This report summarizes the number of safety incidents that have occurred (such as collisions and fires) and the number of security incidents that have occurred in a fixed number of categories. The Non-Major Summary Report form gathers transit incidents that are not reported on the Major Incident Reporting form.

A transit event is considered a non-major *safety* incident when one or more of the following conditions exist:

1. Injuries requiring immediate medical attention away from the scene for one person;
2. Property damage equal to or exceeding $7,500 but less than $25,000; and/or
3. All non-arson fires not qualifying as "Major Incidents."

A non-major *security* incident is recorded when one of the following occurs in the transit environment:

1. *Part I Offenses* that includes forcible rape, robbery, aggravated assault, burglary, larceny/theft, motor vehicle theft, arson;

2. Arrest/citation for *Part II Offenses* that includes vandalism, trespassing, fare evasion, and other assaults;
3. Other security issues that includes bomb threat, bombing, chemical/biological release, cyber incident, hijacking, non-violent civil disturbance, sabotage; or
4. Suicides and attempts.

A portion of the Non-Major Summary Report form found on the NTD website is shown in Figure 2.21.

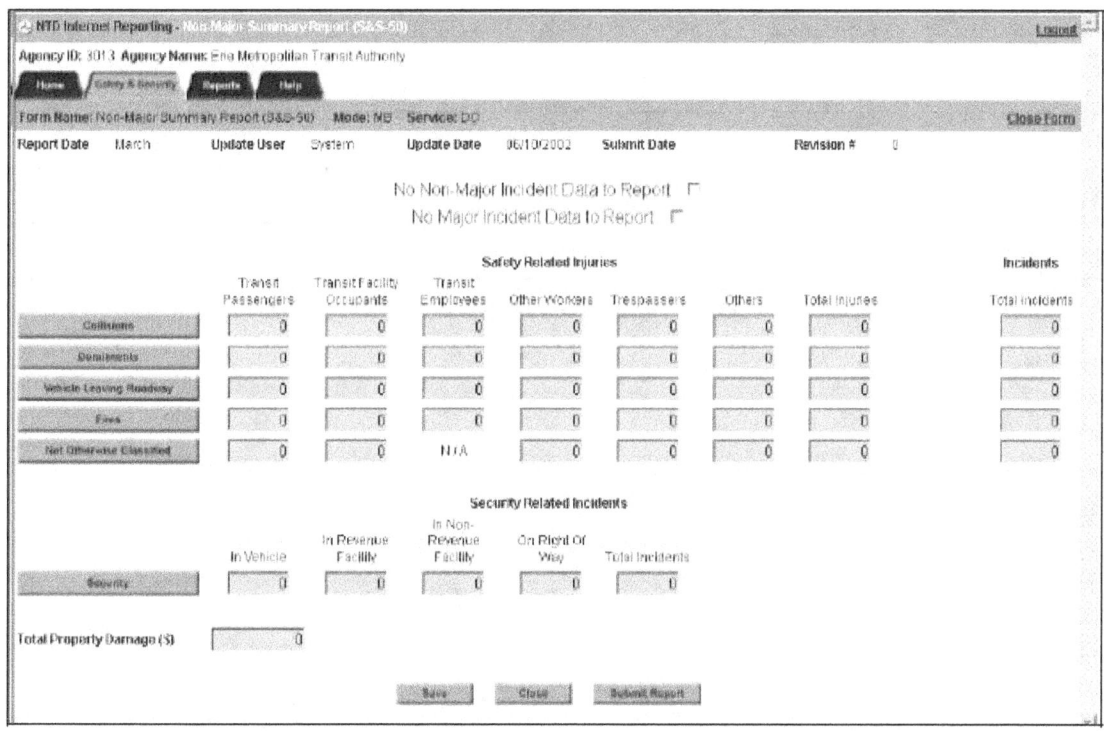

Figure 2.21. Non-Major Summary Report Form from the NTD [FTA, 2002]

2.5.2 Breakdown of Incident Types and Associated Costs

As shown in Figure 2.21, the safety-related incidents in the Non-Major Incident Reporting form are categorized into five types:

A. Collisions
B. Derailments
C. Vehicles Leaving Roadway
D. Fires
E. Not Otherwise Classified

Information on security-related incidents is also gathered using the Non-Major Incident Reporting form. Table 2.5 presents a summary of the non-major transit bus incidents divided by various categories. Figure 2.22 shows the percentage distribution of the non-major incident counts by: non-major safety incident, non-major security incident, and missing information (i.e., records with no information on incident type – non-major safety incident or non-major security incident).

Table 2.5. Breakdown of Non-Major Bus Incidents Recorded in the 2002 NTD[A]

Incident Category	Number of Records	Total Property Damage	Number of Fatalities	Number of Injuries
Non-Major Safety Incident	12,450	$13,238,091	0	7,868
Collisions	*6,932*	*$12,001,697*	*0*	*3,288*
Vehicles Leaving Roadway	*27*	*$159,141*	*0*	*17*
Fires	*136*	*$945,710*	*0*	*21*
Not Otherwise Classified	*5,355*	*$131,543*	*0*	*4,542*
Non-Major Security Incident	8,932	$137,115	1	31
Total	**21,382**	**$13,375,206**	**1**	**7,899**

[A] Final reporting of the 2002 NTD – information as of July 2, 2003.

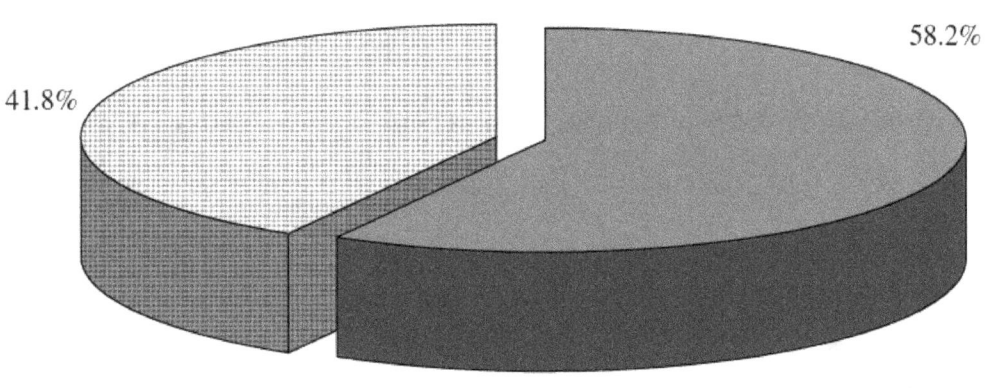

Figure 2.22. Distribution of Non-Major Bus Incidents from the 2002 NTD

As shown in Table 2.5, a total of 21,382 bus-related non-major incident records were filed in the 2002 NTD. A total of 369 transit agencies throughout the United States provided information on non-major incidents in 2002. The number of non-major bus incidents in monthly reports submitted by various transit agencies ranged from 0 to 474 non-major incidents.

According to Figure 2.22, approximately 40 percent of the non-major incidents reported in the 2002 NTD are related to security. However, the property damage costs resulting from non-major security incidents are considerably lower compared to the costs resulting from non-major safety incidents.

Table 2.5 also shows that there are 12,450 non-major safety incidents reported in the 2002 NTD, resulting in more than $13 million in property damage. Average property damage cost per non-major safety incident is $1,063. Further breakdown of the average property damage cost by non-major safety incident type (see Figure 2.23) indicates that an average non-major bus collision costs $1,731 in property damage, considerably lower than the average property damage costs for non-major "fire" and "vehicle leaving roadway" incidents.

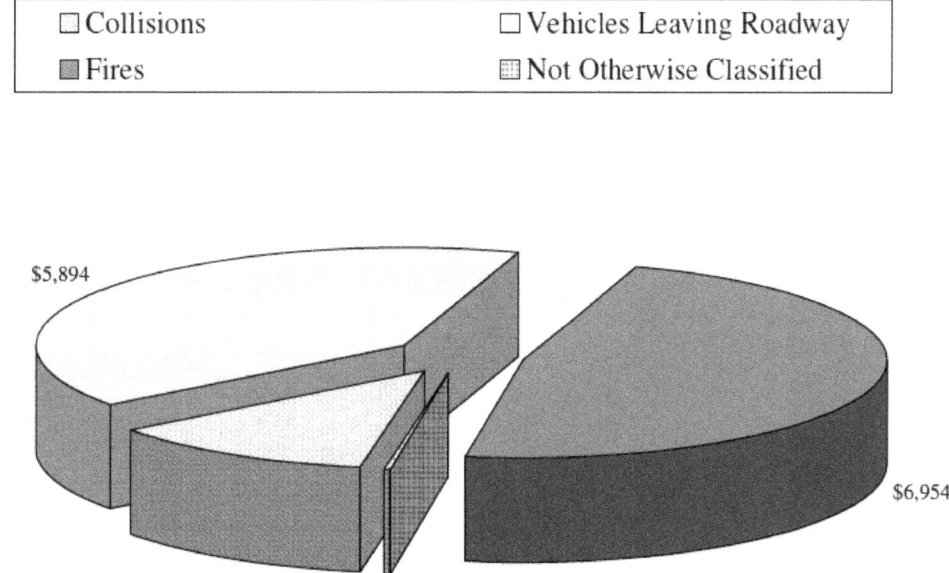

Figure 2.23. Average Property Damage Cost Per Types of Non-Major Safety Incidents

2.5.3 Other Information from the Non-Major Incident Records

Non-major incident records contain limited information compared to the data gathered in major incident records. However, a specific piece of information collected as part of the non-major incident records is of interest for this study – service type of the bus involved in non-major incident (i.e., directly operated versus purchased transportation).

As mentioned previously, *directly operated service* refers to transportation service provided directly by a transit agency while *purchased transportation service* is provided to a public transit agency from a public or private transportation provider.

Table 2.6 shows a summary of non-major incidents by type of service. Of the 12,450 reported non-major incidents, 12,012 (96.5 percent) records involved directly operated service buses and 438 (3.5 percent) events involved purchased transportation. A similar trend is observed when we look only at the 6,932 bus collision records within the non-major safety incident category – a majority of these accidents (97.6 percent) belong to directly operated service buses.

Table 2.6. Distribution of Non-Major Bus Incidents by Type of Service

Type of Service	Number of Non-Major Incidents	Number of Non-Major *Safety* Incidents – *Collisions Only*
Directly Operated	12,012	6,769
Purchased Transportation	438	163
Total	**12,450**	**6,932**

3. COMPARING RESULTS FROM THE NATIONAL TRANSIT DATABASE TO A TRANSIT ACCIDENT ANALYSIS STUDY CONDUCTED BY PATH

This section will present selected results from a report prepared by the California PATH Program at the University of California titled *Development of Requirement Specifications for Transit Frontal Collision Warning System* [Wang et al., March 2002]. A section of this report documented findings from a bus accident data analysis performed by the PATH's research team. Data analysis results presented in Section 2 of this report will be compared with the information summarized in PATH's report.

3.1 Overview of PATH's Transit Accident Data Analysis Work

3.1.1 Background on PATH's Project

PATH's report summarized findings from their work related to a frontal collision warning system for transit buses. In addition to the California PATH Program, other project partners for this study included the FTA, San Mateo County Transit District, California Department of Transportation, and Gillig Corporation. This project started in January 2000 and was completed in March 2002.

The project report prepared by the PATH research team is divided into three major sections. Section 1 summarizes findings from the accident data analysis the research team has performed. Section 2 provides a detailed description of prototype collision warning systems developed by PATH for field testing of the frontal collision warning system. Section 3 of the report provides preliminary performance requirement specifications for the transit frontal collision warning system, generated as part of their project.

Only selected information from Section 1 of PATH's report will be described in this report because the objective is to compare accident data analysis results reported in the PATH report to findings from the analysis of the 2002 NTD.

3.1.2 Sources of Data for this Study

Since one of the objectives of PATH's project is to understand the cause and consequences of transit frontal collision, their research team reviewed and analyzed transit accident data from several sources.

PATH's team reviewed national bus accident statistics published in *Traffic Safety Facts 2000: A Compilation of Motor Vehicle Crash Data from the Fatality Analysis Reporting System and the General Estimates System* [NHTSA, December 2001]. Information presented in this NHTSA report is generated from data found in the FARS and GES.

Besides reviewing the national statistics, PATH's research team also received and analyzed accident data from 35 transit agencies within the State of California – 3 agencies in the San Francisco Bay Area and 32 transit agencies who are members of California Transit Insurance Pool (CalTIP) and located in various cities throughout California. PATH received 5 years of transit accident data from these 35 California transit agencies – May 1, 1997, to April 30, 2002.

3.2 Key Results Reported from PATH's Transit Accident Data Analysis

This sub-section of the report presents results from PATH's analytical work using the data from the 35 California transit agencies. Comparisons between PATH's results and findings in Section 2 of this report using the 2002 NTD will be made throughout this sub-section.

<u>Comparison of Average Costs Resulted from Bus Collisions.</u> PATH's report presented cost information resulting from bus collisions. The following data are taken from PATH's report [Wang *et al.*, March 2002]:

Table 3.1. Costs of Bus Collision from 35 Transit Agencies in California[A]

Transit Agency Code[B]	Number of Claims[C]	Total Cost	Average Cost Per Claim
Agency I	353	$2,904,763	$8,229
Agency II	1,146	$6,319,107	$5,514
Agency III	358	$997,982	$2,788
Agency IV	261	$1,032,796	$3,957
Total	**2,118**	**$11,254,648**	**$5,314**

[A.] Data taken from Table 1.1 and Appendix IV of the PATH report.
[B.] Agencies I and II are two transit agencies in the San Francisco Bay Area. Agencies III and IV are members of CalTIP. Data for the other 31 transit agencies are found to be inconsistent throughout PATH's report; hence, these data were not shown in the table above.
[C.] Each accident is indicated as a "claim" in the database used by PATH. Each claim involves a bus and another party (e.g., vehicle, passenger on the bus, and stationary object). The cost for each claim is the sum of the following cost categories: "Body Injury," "Property Damage," and "Legal and Other Fees."

Average property damage costs resulting from bus collisions, according to "Major Incident" and "Non-Major Incident" records in the 2002 NTD, are $6,151 and $1,731 respectively. The average cost per accident claim shown in Table 3.1, from the database used in PATH's work, is $5,314. On the surface, the average bus accident cost from PATH's work falls within the range of the average cost figures derived from the 2002 NTD. However, the average cost figures from the 2002 NTD include only cost related to "property damage," whereas the average cost per claim from PATH's work involved costs on "body injury," "property damage," and "legal and other fees." Hence, the difference between cost figures generated from PATH's work and the 2002 NTD is larger than it seems.

<u>Comparison of Percentage Distribution for Different Types of Bus Collisions.</u> Table 3.2 shows the percentage distribution of various bus collision types from the California transit agencies and the 2002 NTD. Values for the 2002 NTD were also presented as Figure 2.9 of this report.

Data from the 35 California transit agencies showed that, on average, most bus collisions occurred on the side of the vehicle at initial point of impact, followed by the front and the rear of the vehicle respectively. Side bus collisions from the California transit agencies are more than double that of the rear bus collisions. The 2002 NTD, on the other hand, shows that collisions occurring at the front, side, and rear of a bus are fairly even.

Table 3.2. Percentage Distribution for Various Types of Bus Collisions

Data Source	Percentage of Bus Collisions by Type				
	Frontal	Side (Sideswipe & Angle)	Rear	Fixed Object/Other	Not Sure
2002 NTD	32.3%	30.3%	29.3%	8.1%	N/A
California Transit Agency[A] (Collision Type by Initial Point of Impact)					
Agency I[B]	31.7%	52.4%	10.2%	N/A	5.7%
Agency II[B]	14.7%	40.2%	6.2%	N/A	38.9%
Agency III[B]	17.3%	45.3%	23.7%	N/A	13.7%
Agency IV[B]	29.9%	34.1%	24.9%	N/A	11.1%
Agency V[B]	25.3%	48.0%	16.1%	N/A	10.6%
CalTIP(30)[B]	22.6%	43.3%	18.9%	N/A	15.2%
Average for CA Data	*23.6%*	*43.9%*	*16.7%*	*N/A*	*15.9%*

[A.] Data taken from Figures 1.2, 1.4, 1.6, 1.8, 1.12, and 1.14 of the PATH report.
[B.] Agencies I, II and V are three transit agencies in the San Francisco Bay Area. Agencies III, IV, and CalTIP(30) are the 32 transit agencies who are members of CalTIP.

The difference in percentage distribution of bus collision types between the California data and the 2002 NTD could reflect the fact that there are some regional factors (e.g., traffic characteristics and environmental conditions) specific to California that caused the side bus collisions to be considerably higher and frontal and rear bus collisions to be lower than the national average. Consequently, transit agencies in California should develop strategies that focus on reducing side bus collisions.

Comparison of Cost Distribution for Different Types of Bus Collisions. The cost distribution for frontal, side, and rear bus collisions is presented in Table 3.3. Statistics for the 2002 NTD were presented earlier in Figure 2.10.

Comparing the percentage distribution in cost between the transit agencies in California versus the 2002 NTD, several interesting observations are noted:

1. Both data sources show that frontal collision has the highest cost distribution (in average), followed by side and rear collisions, respectively; however, the magnitude of the difference from one collision type to the next between the two data sources varies considerably;
2. Average percentages of cost for the side collision from the two data sources are similar, while the 2002 NTD cost percentage for rear collision is approximately four times higher than the California collision cost percentage; and
3. Cost percentage for side collision in California's data is less than the cost percentage for frontal collision, even though more side collisions occurred in California compared to frontal collisions (see Table 3.2).

Statistics from Tables 3.2 and 3.3 suggest that transit agencies in California should put emphasis on developing methods to reduce frontal and side bus collisions because the average percentage distributions for these two collision types are higher than the national numbers reported in the 2002 NTD. Rear bus collisions in California are less severe compared to frontal and side collisions.

Table 3.3. Cost Distribution for Various Types of Bus Collisions

Data Source	Cost (in %) of Bus Collisions by Type				
	Frontal	Side (Sideswipe & Angle)	Rear	Fixed Object/Other	Not Sure
2002 NTD	32.9%	27.6%	23.2%	16.3%	N/A
California Transit Agency[A] (Collision Type by Initial Point of Impact)					
Agency I[B]	70.6%	22.5%	2.5%	N/A	4.4%
Agency II[B]	20.5%	45.0%	0.9%	N/A	33.6%
Agency III[B]	22.4%	17.5%	5.7%	N/A	54.4%
Agency IV[B]	61.4%	19.6%	14.3%	N/A	4.7%
Agency V[B]	59.0%	30.6%	4.3%	N/A	6.1%
CalTIP(30)[B]	42.9%	26.8%	5.9%	N/A	24.4%
Average for CA Data	*46.1%*	*27.0%*	*5.6%*	*N/A*	*21.3%*

[A.] Data taken from Figures 1.2, 1.4, 1.6, 1.8, 1.12, and 1.14 of the PATH's report.
[B.] Agencies I, II and V are three transit agencies in the San Francisco Bay Area. Agencies III, IV, and CalTIP(30) are the 32 transit agencies who are members of CalTIP.

4. UTILIZING TRANSIT INTELLIGENT VEHICLE INITIATIVE TECHNOLOGY TO REDUCE TRANSIT BUS ACCIDENTS

The U.S. DOT's Intelligent Vehicle Initiative (IVI) program is a multi-agency research and development endeavor aimed at accelerating the development, availability, and use of driving assistance and control intervention systems to reduce vehicle accidents. IVI systems' ultimate goal is to help drivers operate vehicles more safely. A "platform" within the IVI program focuses attention on transit vehicles. Several key projects under the transit IVI platform are studying and developing collision warning systems that will help minimize transit accidents, such as those described in Section 2.

Section 4.1 introduces four key collision warning systems projects within the transit IVI platform: frontal collision warning system, side object detection system and side collision warning system, rear impact collision warning system, and integrated collision warning system. These collision warning systems are intended to reduce "imminent crash situations" in the transit operating environment. Section 4.2 will use examples to demonstrate the potential cost savings from deploying transit collision warning systems by reducing frontal, side, and rear collisions.

The ultimate goal for the transit collision warning systems is to provide bus operators with effective and timely warnings regarding potential accidents. Transit travel will become more attractive and more efficient to commuters if the number of transit accidents can be effectively reduced using these collision warning systems.

4.1 Overview of Transit Intelligent Vehicle Initiative Technology Being Developed

Brief summaries of the transit collision warning systems being studied under the transit IVI platform are presented in this section. Detailed descriptions of these transit collision warning systems can be found in a report titled *2003 Status Report on Transit Intelligent Vehicle Initiative Studies* [Yang *et al.*, June 2003].

4.1.1 Frontal Collision Warning System

The Federal Transit Administration and its partners from universities and the transit industry are cooperatively developing a frontal collision warning system for transit buses under the transit IVI program. A frontal collision warning system uses sensors to detect obstacles in front of the bus and to determine if these obstacles are potential collision hazards. It then provides the bus operator with warnings, which change in severity as the likelihood of a collision increases. A prototype frontal collision warning system was developed to permit field-testing of different system elements and for validation of the final performance specifications. The prototype includes: sensors that detect the presence of objects; algorithms that identify and interpret potential hazardous targets, determine threat levels, and generate warnings; and a Driver Vehicle Interface that communicates the warning message to the operator. The Driver Vehicle Interface must provide warnings without unduly interfering in the operator's control of the bus.

A prototype frontal collision warning system was developed and deployed on three buses to gather field data and verify that the performance specifications for this system were reasonable and achievable. Researchers for this project are continuing to develop various components of the frontal collision warning system to ensure that it will help reduce both the frequency and severity of frontal collisions with transit buses.

4.1.2 Side Object Detection System and Side Collision Warning System

There are two projects that are examining issues of side collisions with transit buses.

The first project tests a transit-specific object detection system. The prototype object detection system on the buses uses ultrasonic sensors to detect conflicts that are just to the side of a transit bus. Placed intermittently along either side, these sensors provide coverage for detecting objects and pedestrians during close maneuvers, whether in or out of blind spots, and for monitoring lane markers to help the operator negotiate lane changes and turns. When the sensors detect potential obstacles, bus operators are alerted via audible or visible warnings.

A 1-year field operational test (FOT) of the side object detection system project began in April 2001. A primary goal of the FOT was to identify the operation and maintenance issues toward achieving driver acceptance of this technology. Through responses from system evaluation questionnaires, bus operators and maintenance staff provided feedback on the side object detection technology:

- 66 percent reported the side object detection system reinforced safe driving habits;
- 63 percent reported the side object detection system help detect objects in the blind spot; and
- 78 percent said if their concerns about the audible alarm, zone sizes, and sensor placement of the side object detection system could be resolved, the technology would make sense.

Another key FOT objective was to assess economic viability to determine if further installations would be justified. Pre- and post-system installation data of bus accidents and claims were analyzed. Findings suggest that side object detection systems contribute to lower side collision accident and claims rates.

The detailed work for advancing the side object detection system has continued with lessons learned from the 1-year FOT. Five buses were equipped with an enhanced side object detection system for a field evaluation test to address specific technical issues.

In conjunction with the side object detection system study, work is underway to develop an advanced side collision warning system. This study does not presuppose a specific sensor, but considers all components and processors for detecting motion, assessing conflicts, and generating appropriate warnings. Applicable sensor technologies examined in this study included: video detectors, motion detectors, radar, sonar, and laser diodes.

Preliminary performance specifications and testing requirements of a fully functional side collision warning system were drafted. Detailed analysis of the accident data was carried out. Additionally, a test bus equipped with a video camera, laser scanner, and curb detector was used to gather data about the operating environment for transit buses.

The research team for the side collision warning system is updating the performance specifications for such a system in accordance with results of the accident data analysis and information gathered from the test bus. Also, a prototype of the Driver Vehicle Interface for this system is being examined.

4.1.3 Rear Impact Collision Warning System

The primary function of the rear impact collision warning system is to sense the rearward scene of transit buses and provides warnings to following vehicles about potential rear-end collisions. Research shows that many rear impact collisions occur under benign driving conditions, where the bus is traveling in a straight line or is stopped. In addition, driver distraction seems to be a significant contributing factor for rear impact collisions.

The rear impact collision warning system project generated performance specifications for such a system to be used on transit buses. Major tasks of this study include:

- Analyze Crash Data – Data analysis showed that 67 percent of rear-end crashes occurred when buses are stopped in lane; nearly 80 percent of rear impact crashes took place during daylight hours and under no adverse weather conditions; and close to 60 percent of these crashes happened on level roadways.
- Develop Data Collection System – A system consists of a computer and disk storage, an automotive laser radar sensor, a video camera, Global Positioning System, and digital and analog input/output ports for bus signal inputs and control of the warning device was developed for use on test buses.
- Establish Performance Specifications – Preliminary performance specifications and evaluation criteria for the rear impact collision warning system have been established. In addition, a warning algorithm for the system was developed and coded into the data collection system.
- Estimate System Benefits – System deployment benefits of the rear impact collision warning system have been estimated. The approximate annual savings per bus from accident reduction is $9,700, if a rear impact collision warning system can be deployed across an entire bus fleet. This figure provides an upper limit in justifying the implementation and operation of rear impact collision warning systems for transit agencies.
- Conduct Field Validation Test and Data Analysis – Data collection for field validation tests was carried out and two buses were used for the field validation test. Data gathered from the field tests provided valuable information on the potential reduction of rear impact collision and helped the research team to assess the effectiveness of warning algorithms used in this study.

4.1.4 Integrated Collision Warning System

The integrated collision warning system project is aimed at developing a system that is both commercially viable and acceptable to bus operators; the system will be built upon the frontal collision warning system and side collision warning system technologies. A single common interface that can provide warnings on both frontal and side hazards will be designed and implemented. Work related to this project is underway and will be completed by May 2005.

Major tasks to be performed on this integrated collision warning system project include:

- Integrate Side and Frontal Collision Warning Systems into a Unified System – This effort will lead to the development of a unified collision warning system specification and limited prototype operational tests.
- Develop a Driver Vehicle Interface Prototype – Specifications will be developed based on findings from operational tests of a commercial interface, simulation testing by

operators, initial Driver Vehicle Interface design options, and design guidelines generated by human factors research.
- Conduct Operational Testing and Evaluation of Enhanced Commercial Systems – The evaluation test will be carried out using five buses equipped with the newest commercial technology, providing collision warning coverage on the side and the left and right front corners of a transit bus.
- Accelerate Deployment of the Integrated Collision Warning System – The research team is working with transit stakeholders to evaluate market potential in deploying integrated collision warning system. The research team will also incorporate manufacturing perspectives into the design and development process of the system, facilitate discussion on technology transfer, conduct development and cost-benefit tradeoffs research, and further refine specifications for this system with industry.
- Test the Integrated Collision Warning System for Commercialization (*Pending Future Funding*) – The research team will also work with a manufacturer to develop, produce, and install a number of prototype integrated collision warning systems for field operational testing and evaluation.

4.2 Potential of Transit Intelligent Vehicle Initiative Technology

The primary objective of transit collision warning systems is to provide bus operators with effective and timely warnings regarding potential accidents. By reducing the number of potential transit accidents with the help of transit IVI technologies, transit agencies should anticipate significant cost savings. This sub-section presents three *hypothetical examples* to demonstrate the potential payoff of deploying transit IVI technologies.

Examples presented in this sub-section use a combination of actual and hypothetical values in the calculation. "Assumed cost to implement a collision warning system" and "percent of collisions being avoided" are assumed values because *no* reliable figures on these variables are available currently. "Average property damage costs" used in the examples are taken from Table 2.2, Table 2.5, Figure 2.13, and Figure 2.23 of this report, based on the analysis of the 2002 NTD.

It is important to note that only the property damage costs were considered in the three examples. Bus collisions will usually incur other expenses besides property damage. Consequently, actual monetary savings as a result of deploying transit collision warning systems could be higher than the numbers shown in Tables 4.1, 4.2, and 4.3.

Example 1 – Potential Monetary Benefit of Deploying a Transit Frontal Collision Warning System (FCWS). Table 4.1 presents the potential cost savings of deploying the frontal collision warning system, assuming the cost to deploy a system is $7,500 and the system can avoid 85 percent of potential frontal collisions. Combinations of 4 levels of 'Number of Buses with FCWS" (80, 250, 750, and 1200) and 3 levels of "Estimated Number of Frontal Collisions Per Year" (50, 100, and 250) are also listed in Table 4.1. (Note. Cost presented is an estimate based on prototype system developed. System cost should decrease over time.)

According to Table 4.1, if a transit agency plans to install 80 frontal collision warning systems and is currently experiencing 40 major frontal collisions and 60 non-major collisions per year, the system deployment cost will be $600,000; monetary savings from system deployment, due to reduction in frontal collisions, will be $301,529. The time required to recoup the system implementation cost is 2.0 years. The potential benefit offered by the frontal collision warning system is very encouraging.

Table 4.1 also shows that if a transit agency wants to install more than 80 frontal collision warning systems, while other factors remain the same, more time is required to recover the deployment cost. Finally, if a transit agency is experiencing a high number of frontal collisions per year, installing a large number of frontal collision warning systems could be justified because time needed to recoup from implementation cost is substantially reduced.

Example 2 – Potential Monetary Benefit of Deploying a Transit Side Object Detection System (SODS). Table 4.2 presents the projected cost savings of deploying the side collision warning system, assuming the cost to deploy a system is $2,500 and the system can help driver to avoid 65 percent of side collisions. Similar to the first example, cost savings and times required to recover from system deployment were calculated for 12 hypothetical scenarios (4 "Number of Buses with SODS" levels X 3 "Estimated Number of Side Collisions Per Year" levels).

Calculated results in Table 4.2 show the anticipated benefits of deploying side object detection system under various hypothetical scenarios. For example, if a transit agency has 40 major side collisions and 60 non-major collisions per year, deploying side object detection system on 250 buses would save that agency more than $200,000 per year and requires approximately 3 years to recoup the deployment cost.

Example 3 – Potential Monetary Benefit of Deploying a Transit Rear Impact Collision Warning System (RICWS). Table 4.3 shows the potential cost savings of adding the rear impact collision warning system on bus, assuming the cost to implement a rear impact collision warning system is $5,000 and such system will prevent 75 percent of potential rear collisions. Once again, cost savings and times required to recover from system deployment were calculated for the same 12 hypothetical scenarios as the first two example.

Table 4.3 shows that if a transit agency has 100 major rear collisions and 150 non-major collisions per year and wants to deploy the rear impact collision warning system on 750 buses, more than 6 years are needed to regain the system implementation cost. However, if the agency only deploys 250 rear impact collision warning systems instead of 750, the time needed to recover the implementation cost is a little more than 2 years.

Through the three examples presented in this sub-section, the various "levels of benefits" that could be realized from the transit collision warning systems were estimated under hypothetical scenarios. As more precise data become available in the future from the ongoing transit IVI projects in regards to comprehensive bus collision cost, system deployment cost, and percent of bus collisions that can be reduced with the help of the collision warning systems, the benefit assessment of the transit collision warning systems will be further refined.

Table 4.1. Potential Cost Saving of Deploying Transit Frontal Collision Warning System

Type of the System =		Frontal Collision Warning System (FCWS)						
Assumed Cost to Implement a FCWS =		$7,500	(This "assumed cost" includes hardware/software for the FCWS and all associated costs.)					
% of Frontal Collisions Being Avoided =		85%	(This estimate implies that the FCWS can effectively avoid 85% of potential frontal collisions.)					

Number of Buses with FCWS	System Cost	Average Property Damage Cost from Frontal Collisions		Estimated Number of Frontal Collisions Per Year		Property Damage Cost from Frontal Collisions			Saving from Collisions Avoided Per Year (Property Damage Only)	Year(s) Needed to Recover from Implementation Cost
		From Major Incident[A]	From Non-Major Incident[B]	Major Incident	Non-Major Incident	Major Incident	Non-Major Incident	Total		
80	$600,000	$6,272	$1,731	20	30	$125,440	$51,930	$177,370	$150,765	4.0
80	$600,000	$6,272	$1,731	40	60	$250,880	$103,860	$354,740	$301,529	2.0
80	$600,000	$6,272	$1,731	100	150	$627,200	$259,650	$886,850	$753,823	0.8
250	$1,875,000	$6,272	$1,731	20	30	$125,440	$51,930	$177,370	$150,765	12.4
250	$1,875,000	$6,272	$1,731	40	60	$250,880	$103,860	$354,740	$301,529	6.2
250	$1,875,000	$6,272	$1,731	100	150	$627,200	$259,650	$886,850	$753,823	2.5
750	$5,625,000	$6,272	$1,731	20	30	$125,440	$51,930	$177,370	$150,765	37.3
750	$5,625,000	$6,272	$1,731	40	60	$250,880	$103,860	$354,740	$301,529	18.7
750	$5,625,000	$6,272	$1,731	100	150	$627,200	$259,650	$886,850	$753,823	7.5
1,200	$9,000,000	$6,272	$1,731	20	30	$125,440	$51,930	$177,370	$150,765	59.7
1,200	$9,000,000	$6,272	$1,731	40	60	$250,880	$103,860	$354,740	$301,529	29.8
1,200	$9,000,000	$6,272	$1,731	100	150	$627,200	$259,650	$886,850	$753,823	11.9

[A]. Data taken from Table 2.2 and Figure 2.13 of this report.
[B]. Data taken from Table 2.5 and Figure 2.23 of this report.
Note: The relationship between "Number of Buses with FCWS," "Estimated Number of Frontal Collisions Per Year," and "Year(s) Needed to Recover from Implementation Cost" is depicted graphically in Figure A.1.

Table 4.2. Potential Cost Saving of Deploying Transit Side Object Detection System

Type of the System =		Side Object Detection System (SODS)						
Assumed Cost to Implement a SODS =		$2,500	(This "assumed cost" includes hardware/software for the SODS and all associated costs.)					
% of Side Collisions Being Avoided[A] =		65%	(This estimate implies that the SODS can effectively avoid 65% of potential side collisions.)					

Number of Buses with SODS	System Cost	Average Property Damage Cost from Side Collisions		Estimated Number of Side Collisions Per Year		Property Damage Cost from Side Collisions			Saving from Collisions Avoided Per Year (Property Damage Only)	Year(s) Needed to Recover from Implementation Cost
		From Major Incident[B]	From Non-Major Incident[C]	Major Incident	Non-Major Incident	Major Incident	Non-Major Incident	Total		
80	$200,000	$5,597	$1,731	20	30	$111,940	$51,930	$163,870	$106,516	1.9
80	$200,000	$5,597	$1,731	40	60	$223,880	$103,860	$327,740	$213,031	0.9
80	$200,000	$5,597	$1,731	100	150	$559,700	$259,650	$819,350	$532,578	0.4
250	$625,000	$5,597	$1,731	20	30	$111,940	$51,930	$163,870	$106,516	5.9
250	$625,000	$5,597	$1,731	40	60	$223,880	$103,860	$327,740	$213,031	2.9
250	$625,000	$5,597	$1,731	100	150	$559,700	$259,650	$819,350	$532,578	1.2
750	$1,875,000	$5,597	$1,731	20	30	$111,940	$51,930	$163,870	$106,516	17.6
750	$1,875,000	$5,597	$1,731	40	60	$223,880	$103,860	$327,740	$213,031	8.8
750	$1,875,000	$5,597	$1,731	100	150	$559,700	$259,650	$819,350	$532,578	3.5
1,200	$3,000,000	$5,597	$1,731	20	30	$111,940	$51,930	$163,870	$106,516	28.2
1,200	$3,000,000	$5,597	$1,731	40	60	$223,880	$103,860	$327,740	$213,031	14.1
1,200	$3,000,000	$5,597	$1,731	100	150	$559,700	$259,650	$819,350	$532,578	5.6

[A] 'Side Collision" = " Sideswipe" and "Angle Collision" from the 2002 NTD
[B] Data taken from Table 2.2 and Figure 2.13 of this report.
[C] Data taken from Table 2.5 and Figure 2.23 of this report.

Note: The relationship between 'Number of Buses with SODS,"'Estimated Number of Side Collisions Per Year," and "Year(s) Needed to Recove r from Implementation Cost" is depicted graphically in Figure A.2.

Table 4.3. Potential Cost Saving of Deploying Transit Rear Impact Collision Warning System

Type of the System =	Rear Impact Collision Warning System (RICWS)
Assumed Cost to Implement a RICWS =	$5,000 (This "assumed cost" includes hardware/software for the RICWS and all associated costs.)
% of Rear Collisions Being Avoided =	75% (This estimate implies that the RICWS can effectively avoid 75% of potential rear collisions.)

Number of Buses with RICWS	System Cost	Average Property Damage Cost from Rear Collisions		Estimated Number of Rear Collisions Per Year		Property Damage Cost from Rear Collisions			Saving from Collisions Avoided Per Year (Property Damage Only)	Year(s) Needed to Recover from Implementation Cost
		From Major Incident[A]	From Non-Major Incident[B]	Major Incident	Non-Major Incident	Major Incident	Non-Major Incident	Total		
80	$400,000	$4,864	$1,731	20	30	$97,280	$51,930	$149,210	$111,908	3.6
80	$400,000	$4,864	$1,731	40	60	$194,560	$103,860	$298,420	$223,815	1.8
80	$400,000	$4,864	$1,731	100	150	$486,400	$259,650	$746,050	$559,538	0.7
250	$1,250,000	$4,864	$1,731	20	30	$97,280	$51,930	$149,210	$111,908	11.2
250	$1,250,000	$4,864	$1,731	40	60	$194,560	$103,860	$298,420	$223,815	5.6
250	$1,250,000	$4,864	$1,731	100	150	$486,400	$259,650	$746,050	$559,538	2.2
750	$3,750,000	$4,864	$1,731	20	30	$97,280	$51,930	$149,210	$111,908	33.5
750	$3,750,000	$4,864	$1,731	40	60	$194,560	$103,860	$298,420	$223,815	16.8
750	$3,750,000	$4,864	$1,731	100	150	$486,400	$259,650	$746,050	$559,538	6.7
1,200	$6,000,000	$4,864	$1,731	20	30	$97,280	$51,930	$149,210	$111,908	53.6
1,200	$6,000,000	$4,864	$1,731	40	60	$194,560	$103,860	$298,420	$223,815	26.8
1,200	$6,000,000	$4,864	$1,731	100	150	$486,400	$259,650	$746,050	$559,538	10.7

A. Data taken from Table 2.2 and Figure 2.13 of this report.
B. Data taken from Table 2.5 and Figure 2.23 of this report.
Note: The relationship between "Number of Buses with RICWS," "Estimated Number of Rear Collisions Per Year," and "Year(s) Needed to Recover from Implementation Cost" is depicted graphically in Figure A.3.

4.3 Other Bus Accident Costs Besides the Property Damage Cost

The only bus accident cost information available from the 2002 NTD is the "property damage." Hence, the three examples presented in Section 4.2 include only the "property damage cost" in the cost calculation. However, bus accidents usually involve other costs beside property damage cost. Examples of other tangible bus accident costs include:

- Medical costs to treat employees and passengers injured from bus accidents;
- Workforce adjustment costs such as drug tests and job retraining for bus operators involved in accidents, assignment and training replacement bus operators, and the administrative efforts to process work changes;
- Emergency response costs for police officers, fire fighters, emergency personnel; and
- Insurance administration and legal/court costs such as effort to process insurance claims, legal fees associated with accident litigation, and payment/settlement fees related to accidents.

In addition to the tangible costs listed above, there could be "intangible" costs associated with the bus accidents, such as travel delay cost (for commuters who get caught in traffic congestion), cost of lives from fatalities caused by bus accidents, and negative perception toward transit buses.

Consequently, the "true cost" of a bus accident is higher than the cost estimates presented in the three examples. If "property damage cost" accounts for half of tangible cost from a typical bus accident and other costs listed above account for the other half of the tangible cost, then all values shown in the "Saving from Collisions Avoided Per Year" column of Tables 4.1, 4.2, and 4.3 will be doubled. In addition, all numbers shown in the "Year(s) Needed to Recover from Implementation Cost" column of Tables 4.1 to 4.3 will be reduced by half. Therefore, the return-on-investment from implementing transit IVI technologies has the potential to be even greater than what was presented in the three examples above. Reducing the number of bus accidents will result in significant monetary savings for transit agencies; in addition, it could prompt the public to view bus as a safer mode of travel that is equipped with cutting-edge technology, thereby promoting transit's image, growth, and ridership.

5. CONCLUSION

5.1 Foreseeable Impact of Transit Intelligent Vehicle Initiative Technology

The 2002 NTD provided valuable information regarding bus collisions. Information presented in Section 2 gave detailed information on the property damage costs, injuries, and fatalities related to various types of bus collisions. In addition, effects of various environmental and roadway characteristics on bus collisions were presented. The FTA and its transit IVI project partners will utilize the bus collision information presented in this report as a guide to plan for future research and funding direction and to continue to develop and refine various types of transit collision warning systems. The transit IVI technologies currently being studied and developed have great potential to dramatically reduce the number of bus collisions and further improve the safety of transit travel.

Examples presented in Section 4.2 illustrated the monetary savings that could be realized by deploying transit IVI technologies on transit vehicles to effectively reduce bus collisions. However, the impact of reduced bus collision rate is beyond just the monetary savings. If the number of bus collisions can be effectively reduced, then:

- Injuries and fatalities from bus accidents will also decrease;
- Transit operators will feel less stressful and have more confidence to drive bus in and around congested urban environments;
- Transit agencies could invest the cost savings in other aspects of transit operation and maintenance;
- Additional human resources at transit agencies could be available to address other needs instead of dealing with frequent accident issues;
- Traffic congestion and delays caused by bus accidents will be alleviated; and
- The public will view the bus as a safer mode of travel that is equipped with cutting-edge technology, thereby promoting transit's image, growth, and ridership.

5.2 Suggestions for Implementing the "Suitable" Advanced Technology for Your Transit Fleet

Transit IVI technologies have great potential to reduce bus collisions and further improve the safety aspect of bus travel. However, the initial system deployment cost could be considerably expensive, as illustrated in the hypothetical scenarios shown in the examples in Section 4.2. If a transit agency plans to implement the collision warning system on its buses, a few general guidelines should be followed to maximize the benefits of such advanced technology:

- Develop a strategic technology implementation plan by studying and learning the capability and intended functions of various transit collision warning systems;
- Identify the specific bus collision challenge(s) that need(s) to be addressed; for example, Section 3 showed that California's transit agencies should focus their efforts on reducing side and frontal bus collisions;
- Implement specific collision warning system(s) that would combat the particular problem area(s) of bus collisions the agency faces; and

- Develop a ranking of "bus collision probability" by service regions in the transit agency and install collision warning systems on buses that serve on regions with high bus collision probability instead of the entire bus fleet (as shown from examples in Section 4, the cost recovery time from system deployment will be longer as more buses are install with the collision warning system).

REFERENCES

National Highway Traffic Safety Administration (December 2001). *Traffic Safety Facts 2000: A Compilation of Motor Vehicle Crash Data from the Fatality Analysis Reporting System and the General Estimates System* (DOT HS 809 337). Washington, D.C.: U.S. Department of Transportation.

National Highway Traffic Safety Administration (December 2002). *Traffic Safety Facts 2001: A Compilation of Motor Vehicle Crash Data from the Fatality Analysis Reporting System and the General Estimates System* (DOT HS 809 484). Washington, D.C.: U.S. Department of Transportation.

Powers, G. (March 2002). *Transit Safety & Security Statistics and Analysis: 2000 Annual Report* (FTA-MA-26-5011-02-1). Washington, D.C.: U.S. Department of Transportation.

Federal Transit Administration (2002). *National Transit Database 2002 Reporting Manual* (http://www.ntdprogram.com). Washington, D.C.: U.S. Department of Transportation.

Wang, X., Lins, J., Chan, C.-Y., Johnston, S., Zhou, K., Steinfeld, A., Hanson, M., and Zhang, W.-B. (March 2002). *Development of Requirement Specifications for Transit Frontal Collision Warning System.* Berkeley, CA: California PATH Program, University of California.

Yang, C. Y. D., Cronin, B. P., Meltzer, N. R., and Zirker, M. E. (June 2003). *2003 Status Report on Transit Intelligent Vehicle Initiative Studies* (FHWA-OP-03-092/FTA-TRI-11-2003.1). Washington, D.C.: U.S. Department of Transportation.

APPENDIX A. RELATIONSHIP PLOTS ON NUMBER OF COLLISION WARNING SYSTEMS, ANNUAL BUS INCIDENT COUNTS, AND COST RECOVERY TIME FROM SYSTEM DEPLOYMENT

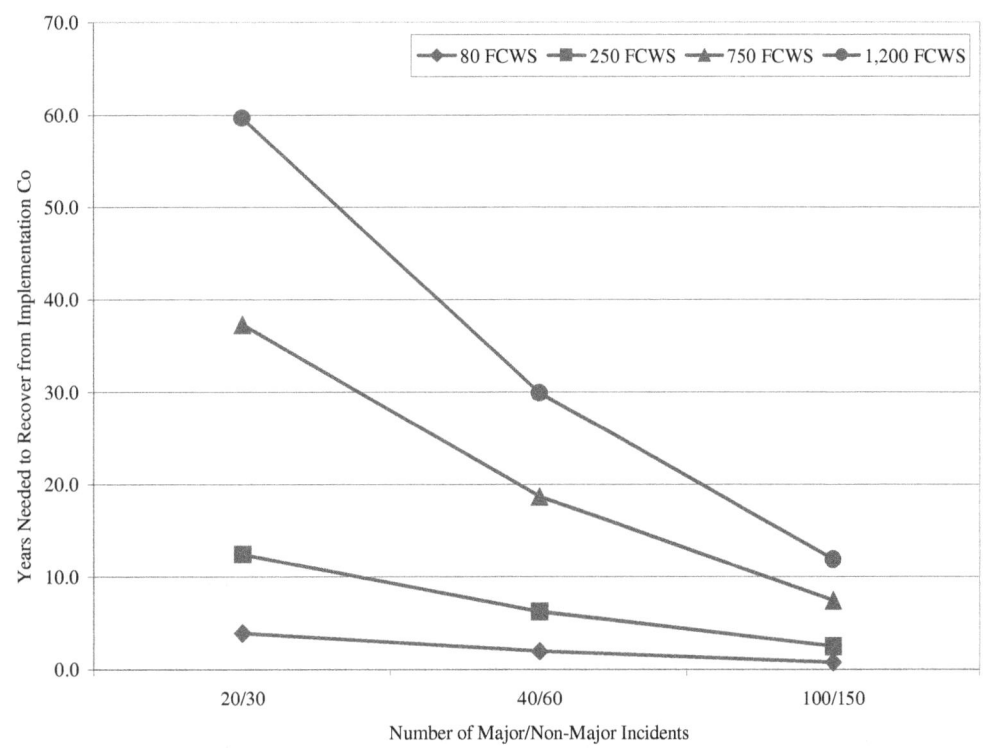

Figure A.1. Time Required to Recoup from Implementation of the Frontal Collision Warning System

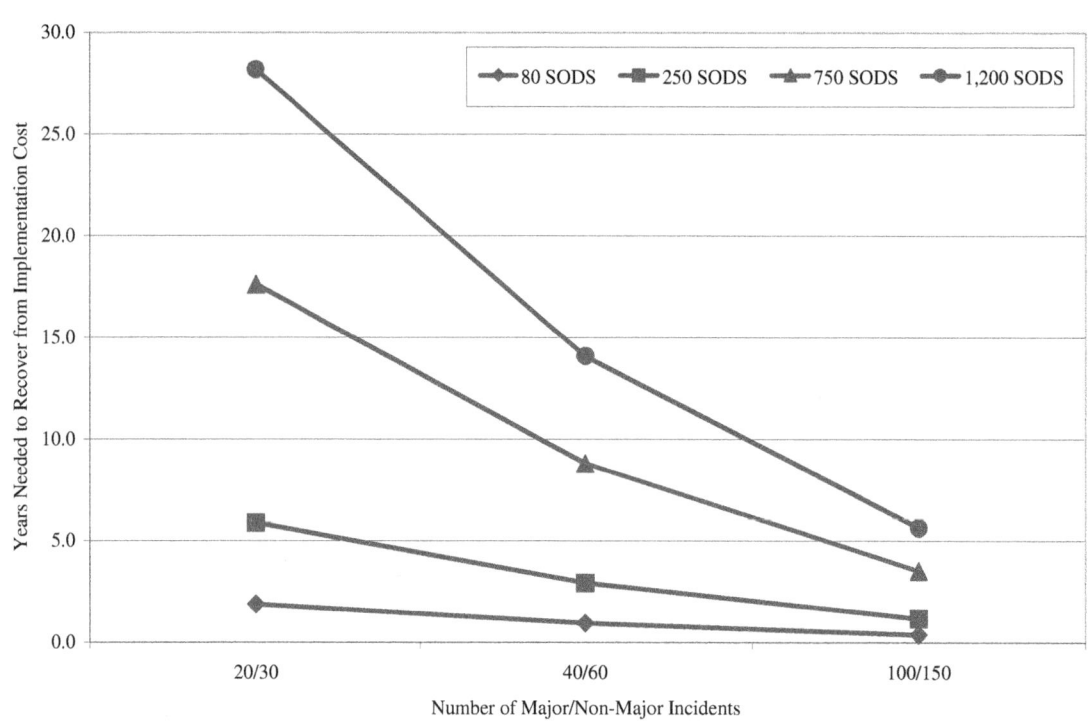

Figure A.2. Time Required to Recoup from Implementation of the Side Object Detection System

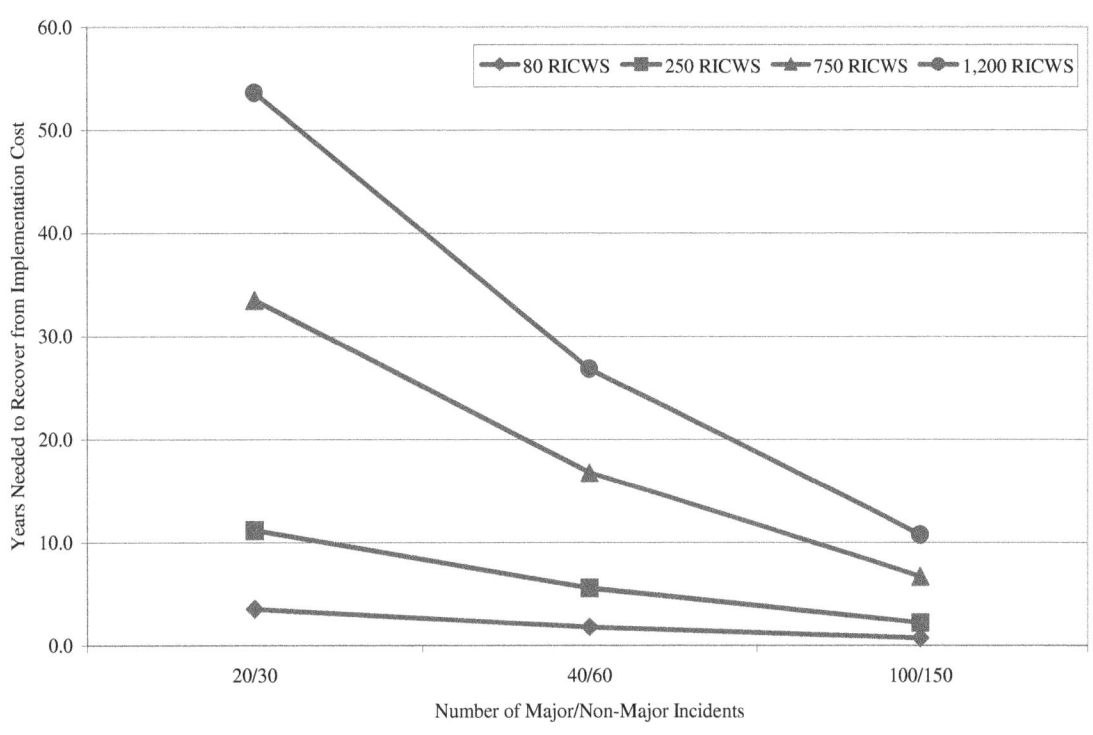

Figure A.3. Time Required to Recoup from Implementation of the Rear Impact Collision Warning System

www.ingramcontent.com/pod-product-compliance
Lightning Source LLC
Chambersburg PA
CBHW081859170526
45167CB00007B/3076